www.EffortlessMath.com

... So Much More Online!

- ✓ FREE Math lessons
- ✓ More Math learning books!
- ✓ Mathematics Worksheets
- ✓ Online Math Tutors

Need a PDF version of this book?

Visit www.EffortlessMath.com

Or send email to: info@EffortlessMath.com

SSAT Middle Level Mathematics Prep 2019

A Comprehensive Review and Ultimate Guide to the SSAT Middle Level Math Test

By

Reza Nazari & Ava Ross

All inquiries should be addressed to:

info@effortlessMath.com

www.EffortlessMath.com

ISBN-13: 978-1-970036-03-9

ISBN-10: 1-970036-03-6

Published by: Effortless Math Education

www.EffortlessMath.com

Description

SSAT Middle Level Mathematics Prep 2019 provides students with the confidence and math skills they need to succeed on the SSAT Middle Level Math, building a solid foundation of basic Math topics with abundant exercises for each topic. It is designed to address the needs of SSAT Middle Level test takers who must have a working knowledge of basic Math.

This comprehensive book with over 2,500 sample questions and 2 complete SSAT Middle Level tests is all you need to fully prepare for the SSAT Middle Level Math. It will help you learn everything you need to ace the math section of the SSAT Middle Level.

There are more than 2,500 Math problems with answers in this book.

Effortless Math unique study program provides you with an in-depth focus on the math portion of the exam, helping you master the math skills that students find the most troublesome.
This book contains most common sample questions that are most likely to appear in the mathematics section of the SSAT Middle Level.

Inside the pages of this comprehensive SSAT Middle Level Math book, students can learn basic math operations in a structured manner with a complete study program to help them understand essential math skills. It also has many exciting features, including:

- Dynamic design and easy-to-follow activities
- A fun, interactive and concrete learning process
- Targeted, skill-building practices
- Fun exercises that build confidence
- Math topics are grouped by category, so you can focus on the topics you struggle on
- All solutions for the exercises are included, so you will always find the answers
- 2 Complete SSAT Middle Level Math Practice Tests that reflect the format and question types on SSAT Middle Level

SSAT Middle Level Mathematics 2019 is an incredibly useful tool for those who want to review all topics being covered on the SSAT Middle Level test. It efficiently and effectively reinforces learning outcomes through engaging questions and repeated practice, helping you to quickly master basic Math skills.

About the Author

Reza Nazari is the author of more than 100 Math learning books including:
– **Math and Critical Thinking Challenges:** For the Middle and High School Student
– **GRE Math in 30 Days**
– **ASVAB Math Workbook 2018 – 2019**
– **Effortless Math Education Workbooks**
– **and many more Mathematics books ...**

Reza is also an experienced Math instructor and a test–prep expert who has been tutoring students since 2008. Reza is the founder of Effortless Math Education, a tutoring company that has helped many students raise their standardized test scores—and attend the colleges of their dreams. Reza provides an individualized custom learning plan and the personalized attention that makes a difference in how students view math.

You can contact Reza via email at:
reza@EffortlessMath.com

Find Reza's professional profile at:
goo.gl/zoC9rJ

Contents

Chapter 1: Arithmetic

Topics that you'll learn in this chapter:

- ✓ Simplifying Fractions
- ✓ Adding and Subtracting Fractions
- ✓ Multiplying and Dividing Fractions
- ✓ Adding Mixed Numbers
- ✓ Subtract Mixed Numbers
- ✓ Multiplying Mixed Numbers
- ✓ Dividing Mixed Numbers
- ✓ Comparing Decimals
- ✓ Rounding Decimals
- ✓ Adding and Subtracting Decimals
- ✓ Multiplying and Dividing Decimals
- ✓ Converting Between Fractions, Decimals and Mixed Numbers
- ✓ Factoring Numbers
- ✓ Greatest Common Factor
- ✓ Least Common Multiple
- ✓ Divisibility Rules

Simplifying Fractions

Helpful Hints

– Evenly divide both the top and bottom of the fraction by 2, 3, 5, 7, ... etc.
– Continue until you can't go any further.

Example:

$\frac{4}{12} = \frac{2}{6} = \frac{1}{3}$

Simplify the fractions.

1) $\frac{22}{36}$

2) $\frac{8}{10}$

3) $\frac{12}{18}$

4) $\frac{6}{8}$

5) $\frac{13}{39}$

6) $\frac{5}{20}$

7) $\frac{16}{36}$

8) $\frac{18}{36}$

9) $\frac{20}{50}$

10) $\frac{6}{54}$

11) $\frac{45}{81}$

12) $\frac{21}{28}$

13) $\frac{35}{56}$

14) $\frac{52}{64}$

15) $\frac{13}{65}$

16) $\frac{44}{77}$

17) $\frac{21}{42}$

18) $\frac{15}{36}$

19) $\frac{9}{24}$

20) $\frac{20}{80}$

21) $\frac{25}{45}$

Adding and Subtracting Fractions

Helpful Hints

– For "like" fractions (fractions with the same denominator), add or subtract the numerators and write the answer over the common denominator.

– Find equivalent fractions with the same denominator before you can add or subtract fractions with different denominators.

– Adding and Subtracting with the same denominator:

$$\frac{a}{b}+\frac{c}{b}=\frac{a+c}{b}$$

$$\frac{a}{b}-\frac{c}{b}=\frac{a-c}{b}$$

– Adding and Subtracting fractions with different denominators:

$$\frac{a}{b}+\frac{c}{d}=\frac{ad+cb}{bd}$$

$$\frac{a}{b}-\frac{c}{d}=\frac{ad-cb}{bd}$$

✍ Add fractions.

1) $\frac{2}{3}+\frac{1}{2}$

4) $\frac{7}{4}+\frac{5}{9}$

7) $\frac{3}{4}+\frac{2}{5}$

2) $\frac{3}{5}+\frac{1}{3}$

5) $\frac{2}{5}+\frac{1}{5}$

8) $\frac{2}{3}+\frac{1}{5}$

3) $\frac{5}{6}+\frac{1}{2}$

6) $\frac{3}{7}+\frac{1}{2}$

9) $\frac{16}{25}+\frac{3}{5}$

✍ Subtract fractions.

10) $\frac{4}{5}-\frac{2}{5}$

13) $\frac{8}{9}-\frac{3}{5}$

16) $\frac{3}{4}-\frac{13}{18}$

11) $\frac{3}{5}-\frac{2}{7}$

14) $\frac{3}{7}-\frac{3}{14}$

17) $\frac{5}{8}-\frac{2}{5}$

12) $\frac{1}{2}-\frac{1}{3}$

15) $\frac{4}{15}-\frac{1}{10}$

18) $\frac{1}{2}-\frac{1}{9}$

Multiplying and Dividing Fractions

Helpful Hints		
	– **Multiplying fractions:** multiply the top numbers and multiply the bottom numbers. – **Dividing fractions:** Keep, Change, Flip Keep first fraction, change division sign to multiplication, and flip the numerator and denominator of the second fraction. Then, solve!	**Example:** $\frac{a}{b} \times \frac{c}{d} = \frac{a \times c}{b \times d}$ $\frac{a}{b} \div \frac{c}{d} = \frac{a}{b} \times \frac{d}{c} = \frac{ad}{bc}$

Multiplying fractions. Then simplify.

1) $\frac{1}{5} \times \frac{2}{3}$

4) $\frac{3}{8} \times \frac{1}{3}$

7) $\frac{2}{3} \times \frac{3}{8}$

2) $\frac{3}{4} \times \frac{2}{3}$

5) $\frac{3}{5} \times \frac{2}{5}$

8) $\frac{1}{4} \times \frac{1}{3}$

3) $\frac{2}{5} \times \frac{3}{7}$

6) $\frac{7}{9} \times \frac{1}{3}$

9) $\frac{5}{7} \times \frac{7}{12}$

Dividing fractions.

10) $\frac{2}{9} \div \frac{1}{4}$

13) $\frac{11}{14} \div \frac{1}{10}$

16) $\frac{3}{5} \div \frac{1}{5}$

11) $\frac{1}{2} \div \frac{1}{3}$

14) $\frac{3}{5} \div \frac{5}{9}$

17) $\frac{12}{21} \div \frac{3}{7}$

12) $\frac{6}{11} \div \frac{3}{4}$

15) $\frac{1}{2} \div \frac{1}{2}$

18) $\frac{5}{14} \div \frac{9}{10}$

Adding Mixed Numbers

Helpful Hints	Use the following steps for both adding and subtracting mixed numbers.	**Example:**
	– Find the Least Common Denominator (LCD) – Find the equivalent fractions for each mixed number. – Add fractions after finding common denominator. – Write your answer in lowest terms.	$1\frac{3}{4} + 2\frac{3}{8} = 4\frac{1}{8}$

Add.

1) $4\frac{1}{2} + 5\frac{1}{2}$

2) $2\frac{3}{8} + 3\frac{1}{8}$

3) $6\frac{1}{5} + 3\frac{2}{5}$

4) $1\frac{1}{3} + 2\frac{2}{3}$

5) $5\frac{1}{6} + 5\frac{1}{2}$

6) $3\frac{1}{3} + 1\frac{1}{3}$

7) $1\frac{10}{11} + 1\frac{1}{3}$

8) $2\frac{3}{6} + 1\frac{1}{2}$

9) $5\frac{3}{5} + 5\frac{1}{5}$

10) $7 + \frac{1}{5}$

11) $1\frac{5}{7} + \frac{1}{3}$

12) $2\frac{1}{4} + 1\frac{1}{2}$

Subtract Mixed Numbers

Helpful Hints

Use the following steps for both adding and subtracting mixed numbers.

Find the Least Common Denominator (LCD)
– Find the equivalent fractions for each mixed number.
– Add or subtract fractions after finding common denominator.
– Write your answer in lowest terms.

Example:

$5\frac{2}{3} - 3\frac{2}{7} = 2\frac{8}{21}$

Subtract.

1) $4\frac{1}{2} - 3\frac{1}{2}$

2) $3\frac{3}{8} - 3\frac{1}{8}$

3) $6\frac{3}{5} - 5\frac{1}{5}$

4) $2\frac{1}{3} - 1\frac{2}{3}$

5) $6\frac{1}{6} - 5\frac{1}{2}$

6) $3\frac{1}{3} - 1\frac{1}{3}$

7) $2\frac{10}{11} - 1\frac{1}{3}$

8) $2\frac{1}{2} - 1\frac{1}{2}$

9) $6\frac{3}{5} - 2\frac{1}{5}$

10) $7\frac{2}{5} - 1\frac{1}{5}$

11) $2\frac{5}{7} - 1\frac{1}{3}$

12) $2\frac{1}{4} - 1\frac{1}{2}$

Multiplying Mixed Numbers

Helpful Hints

1- Convert the mixed numbers to improper fractions.
2- Multiply fractions and simplify if necessary.

$$a\frac{c}{b} = a + \frac{c}{b} = \frac{ab + c}{b}$$

Example:

$2\frac{1}{3} \times 5\frac{3}{7} =$

$\frac{7}{3} \times \frac{38}{7} = \frac{38}{3} = 12\frac{2}{3}$

Find each product.

1) $1\frac{2}{3} \times 1\frac{1}{4}$

2) $1\frac{3}{5} \times 1\frac{2}{3}$

3) $1\frac{2}{3} \times 3\frac{2}{7}$

4) $4\frac{1}{8} \times 1\frac{2}{5}$

5) $2\frac{2}{5} \times 3\frac{1}{5}$

6) $1\frac{1}{3} \times 1\frac{2}{3}$

7) $1\frac{5}{8} \times 2\frac{1}{2}$

8) $3\frac{2}{5} \times 2\frac{1}{5}$

9) $2\frac{2}{3} \times 4\frac{1}{4}$

10) $2\frac{3}{5} \times 1\frac{2}{4}$

11) $1\frac{1}{3} \times 1\frac{1}{4}$

12) $3\frac{2}{5} \times 1\frac{1}{5}$

Dividing Mixed Numbers

Helpful Hints

1- Convert the mixed numbers to improper fractions.
2- Divide fractions and simplify if necessary.

$$a\frac{c}{b} = a + \frac{c}{b} = \frac{ab + c}{b}$$

Example:

$2\frac{1}{3} \div 5\frac{3}{7} =$

$\frac{7}{3} \div \frac{38}{7} = \frac{7}{3} \times \frac{7}{38} = \frac{49}{108}$

Find each quotient.

1) $2\frac{1}{5} \div 2\frac{1}{2}$

2) $2\frac{3}{5} \div 1\frac{1}{3}$

3) $3\frac{1}{6} \div 4\frac{2}{3}$

4) $1\frac{2}{3} \div 3\frac{1}{3}$

5) $4\frac{1}{8} \div 2\frac{2}{4}$

6) $3\frac{1}{2} \div 2\frac{3}{5}$

7) $3\frac{5}{9} \div 1\frac{2}{5}$

8) $2\frac{2}{7} \div 1\frac{1}{2}$

9) $3\frac{1}{5} \div 1\frac{1}{2}$

10) $4\frac{3}{5} \div 2\frac{1}{3}$

11) $6\frac{1}{6} \div 1\frac{2}{3}$

12) $2\frac{2}{3} \div 1\frac{1}{3}$

Comparing Decimals

Helpful Hints

- **Decimals:** is a fraction written in a special form. For example, instead of writing $\frac{1}{2}$ you can write 0.5.
- **For comparing:**
  Equal to =

  Less than <

  Greater than >

  Greater than or equal ≥

  Less than or equal ≤

Example:

2.67 > 0.267

Write the correct comparison symbol (>, < or =).

1) 1.25 2.3
2) 0.5 0.23
3) 3.2 3.2
4) 4.58 45.8
5) 2.75 0.275
6) 5.2 5
7) 3.1 0.31
8) 6.33 0.733
9) 8 0.8
10) 4.56 0.456
11) 1.12 1.14
12) 2.77 2.78
13) 6.08 6.11
14) 1.11 0.211
15) 2.6 2.55
16) 1.24 1.25
17) 5.52 0.552
18) 0.33 0.033
19) 14.4 14.4
20) 0.05 0.50
21) 0.59 0.7
22) 0.5 0.05
23) 0.90 0.9
24) 0.27 0.4

Rounding Decimals

Helpful Hints

We can round decimals to a certain accuracy or number of decimal places. This is used to make calculation easier to do and results easier to understand, when exact values are not too important.

First, you'll need to remember your place values:

12.4567

1: tens 2: ones 4: tenths

5: hundredths 6: thousandths 7: ten thousandths

Example:

$\underline{6}.37 = 6$

Round each decimal number to the nearest place indicated.

1) $0.\underline{2}3$
2) $4.\underline{0}4$
3) $5.\underline{6}23$
4) $0.\underline{2}66$
5) $\underline{6}.37$
6) $0.\underline{8}8$
7) $8.\underline{2}4$
8) $\underline{7}.0760$
9) $1.6\underline{2}9$
10) $6.\underline{3}959$
11) $\underline{1}.9$
12) $\underline{5}.2167$
13) $5.\underline{8}63$
14) $8.\underline{5}4$
15) $8\underline{0}.69$
16) $6\underline{5}.85$
17) $70.\underline{7}8$
18) $61\underline{5}.755$
19) $1\underline{6}.4$
20) $9\underline{5}.81$
21) $\underline{2}.408$
22) $7\underline{6}.3$
23) $116.\underline{5}14$
24) $8.\underline{0}6$

Adding and Subtracting Decimals

Helpful Hints	1– Line up the numbers. 2– Add zeros to have same number of digits for both numbers. 3– Add or Subtract using column addition or subtraction.	Example: $\begin{array}{r} 16.18 \\ -\ 13.45 \\ \hline 2.73 \end{array}$

Add and subtract decimals.

1) $\begin{array}{r} 15.14 \\ -\ 12.18 \\ \hline \end{array}$ ______

3) $\begin{array}{r} 82.56 \\ +\ 12.28 \\ \hline \end{array}$ ______

5) $\begin{array}{r} 90.37 \\ +\ 56.97 \\ \hline \end{array}$ ______

2) $\begin{array}{r} 65.72 \\ +\ 43.67 \\ \hline \end{array}$ ______

4) $\begin{array}{r} 34.18 \\ -\ 23.45 \\ \hline \end{array}$ ______

6) $\begin{array}{r} 45.78 \\ -\ 23.39 \\ \hline \end{array}$ ______

Solve.

7) _____ + 1.3 = 4.8

8) 4.2 + _____ = 11.6

9) 9.9 + _____ = 16

10) 6.9 + _____ = 16.4

11) _____ + 5.1 = 8.6

12) _____ + 7.9 = 15.2

Multiplying and Dividing Decimals

Helpful Hints

For Multiplication:

– Set up and multiply the numbers as you do with whole numbers.

– Count the total number of decimal places in both of the factors.

– Place the decimal point in the product.

For Division:

– If the divisor is not a whole number, move decimal point to right to make it a whole number. Do the same for dividend.

– Divide similar to whole numbers.

Find each product.

1) 4.5×1.6

4) 8.9×9.7

7) 5.7×7.8

2) 7.7×9.9

5) 15.1×12.6

8) 98.20×100

3) 2.6×1.5

6) 6.9×3.3

9) 23.99×1000

Find each quotient.

10) 9.2 ÷ 3.6

11) 27.6 ÷ 3.8

12) 12.6 ÷ 4.7

13) 6.5 ÷ 8.1

14) 1. 4 ÷ 10

15) 3.6 ÷ 100

16) 4.24 ÷ 10

17) 14.6 ÷ 100

18) 1.8 ÷ 1000

Converting Between Fractions, Decimals and Mixed Numbers

Helpful Hints

Fraction to Decimal:

– Divide the top number by the bottom number.

Decimal to Fraction:

– Write decimal over 1.

– Multiply both top and bottom by 10 for every digit on the right side of the decimal point.

– Simplify.

Convert fractions to decimals.

1) $\frac{9}{10}$

2) $\frac{56}{100}$

3) $\frac{3}{4}$

4) $\frac{2}{5}$

5) $\frac{3}{9}$

6) $\frac{40}{50}$

7) $\frac{12}{10}$

8) $\frac{8}{5}$

9) $\frac{69}{10}$

Convert decimal into fraction or mixed numbers.

10) 0.3

11) 4.5

12) 2.5

13) 2.3

14) 0.8

15) 0.25

16) 0.14

17) 0.2

18) 0.08

19) 0.45

20) 2.6

21) 5.2

Factoring Numbers

Helpful Hints		Example:
	- Factoring numbers means to break the numbers into their prime factors.	
	- First few prime numbers: 2, 3, 5, 7, 11, 13, 17, 19	$12 = 2 \times 2 \times 3$

List all positive factors of each number.

1) 68
2) 56
3) 24
4) 40
5) 86
6) 78
7) 50
8) 98
9) 45
10) 26
11) 54
12) 28
13) 55
14) 85
15) 48

List the prime factorization for each number.

16) 50
17) 25
18) 69
19) 21
20) 45
21) 68
22) 26
23) 86
24) 93

Greatest Common Factor

Helpful Hints		Example:
	- List the prime factors of each number. - Multiply common prime factors.	$200 = 2 \times 2 \times 2 \times 5 \times 5$ $60 = 2 \times 2 \times 3 \times 5$ GCF $(200, 60) = 2 \times 2 \times 5 = 20$

Find the GCF for each number pair.

1) 20, 30

2) 4, 14

3) 5, 45

4) 68, 12

5) 5, 12

6) 15, 27

7) 3, 24

8) 34, 6

9) 4, 10

10) 5, 3

11) 6, 16

12) 30, 3

13) 24, 28

14) 70, 10

15) 45, 8

16) 90, 35

17) 78, 34

18) 55, 75

19) 60, 72

20) 100, 78

21) 30, 40

Least Common Multiple

Helpful Hints

- Find the GCF for the two numbers.
- Divide that GCF into either number.
- Take that answer and multiply it by the other number.

Example:

LCM (200, 60):

GCF is 20

$200 \div 20 = 10$

$10 \times 60 = 600$

Find the LCM for each number pair.

1) 4, 14
2) 5, 15
3) 16, 10
4) 4, 34
5) 8, 3
6) 12, 24
7) 9, 18
8) 5, 6
9) 8, 19
10) 9, 21
11) 19, 29
12) 7, 6
13) 25, 6
14) 4, 8
15) 30, 10, 50
16) 18, 36, 27
17) 12, 8, 18
18) 8, 18, 4
19) 26, 20, 30
20) 10, 4, 24
21) 15, 30, 45

Divisibility Rules

Helpful Hints	- Divisibility means that a number can be divided by other numbers evenly.	**Example:** 24 is divisible by 6, because 24 ÷ 6 = 4

Use the divisibility rules to find the factors of each number.

8   <u>2</u> 3 <u>4</u> 5 6 7 <u>8</u> 9 10

1) 16   2 3 4 5 6 7 8 9 10

2) 10   2 3 4 5 6 7 8 9 10

3) 15   2 3 4 5 6 7 8 9 10

4) 28   2 3 4 5 6 7 8 9 10

5) 36   2 3 4 5 6 7 8 9 10

6) 15   2 3 4 5 6 7 8 9 10

7) 27   2 3 4 5 6 7 8 9 10

8) 70   2 3 4 5 6 7 8 9 10

9) 57   2 3 4 5 6 7 8 9 10

10) 102   2 3 4 5 6 7 8 9 10

11) 144   2 3 4 5 6 7 8 9 10

12) 75   2 3 4 5 6 7 8 9 10

Test Preparation

1) Mr. Jones saves $2,500 out of his monthly family income of $55,000. What fractional part of his income does he save?

A. $\frac{1}{22}$

B. $\frac{1}{11}$

C. $\frac{3}{25}$

D. $\frac{2}{15}$

E. $\frac{1}{15}$

2) What is the missing prime factor of number 360?

$360 = 2^3 \times 3^2 \times$ ___

Write your answer in the box below.

3) Which list shows the fractions in order from least to greatest?

$$\frac{2}{3}, \quad \frac{5}{7}, \quad \frac{3}{10}, \quad \frac{1}{2}, \quad \frac{6}{13}$$

A. $\frac{2}{3}, \frac{5}{7}, \frac{3}{10}, \frac{1}{2}, \frac{6}{13}$

B. $\frac{6}{13}, \frac{1}{2}, \frac{2}{3}, \frac{5}{7}, \frac{3}{10}$

C. $\frac{3}{10}, \frac{2}{3}, \frac{5}{7}, \frac{1}{2}, \frac{6}{13}$

D. $\frac{3}{10}, \frac{6}{13}, \frac{1}{2}, \frac{2}{3}, \frac{5}{7}$

E. $\frac{3}{10}, \frac{6}{13}, \frac{2}{3}, \frac{1}{2}, \frac{5}{7}$

4) Which statement about 5 multiplied by $\frac{2}{3}$ is true?

A. The product is between 2 and 3.

B. The product is between 3 and 4

C. The product is more than $\frac{11}{3}$.

D. The product is between $\frac{14}{3}$ and 5.

E. The product is more than 5.

5) Four one – foot rulers can be split among how many users to leave each with $\frac{1}{6}$ of a ruler?

A. 4

B. 6

C. 12

D. 24

E. 36

6) Last week 24,000 fans attended a football match. This week three times as many bought tickets, but one sixth of them cancelled their tickets. How many are attending this week?

A. 48,000

B. 54,000

C. 60,000

D. 72,000

E. 76,000

7) What is the missing prime factor of number 180?

$180 = 2^2 \times 3^2 \times$ ___

A. 2

B. 3

C. 5

D. 6

E. 7

Answers of Worksheets – Chapter 1

Simplifying Fractions

1) $\frac{11}{18}$

2) $\frac{4}{5}$

3) $\frac{2}{3}$

4) $\frac{3}{4}$

5) $\frac{1}{3}$

6) $\frac{1}{4}$

7) $\frac{4}{9}$

8) $\frac{1}{2}$

9) $\frac{2}{5}$

10) $\frac{1}{9}$

11) $\frac{5}{9}$

12) $\frac{3}{4}$

13) $\frac{5}{8}$

14) $\frac{13}{16}$

15) $\frac{1}{5}$

16) $\frac{4}{7}$

17) $\frac{1}{2}$

18) $\frac{5}{12}$

19) $\frac{3}{8}$

20) $\frac{1}{4}$

21) $\frac{5}{9}$

Adding and Subtracting Fractions

1) $\frac{7}{6}$

2) $\frac{14}{15}$

3) $\frac{4}{3}$

4) $\frac{83}{36}$

5) $\frac{3}{5}$

6) $\frac{13}{14}$

7) $\frac{23}{20}$

8) $\frac{13}{15}$

9) $\frac{31}{25}$

10) $\frac{2}{5}$

11) $\frac{11}{35}$

12) $\frac{1}{6}$

13) $\frac{13}{45}$

14) $\frac{3}{14}$

15) $\frac{1}{6}$

16) $\frac{1}{36}$

17) $\frac{9}{40}$

18) $\frac{7}{18}$

Multiplying and Dividing Fractions

1) $\frac{2}{15}$
2) $\frac{1}{2}$
3) $\frac{6}{35}$
4) $\frac{1}{8}$
5) $\frac{6}{25}$
6) $\frac{7}{27}$
7) $\frac{1}{4}$
8) $\frac{1}{12}$
9) $\frac{5}{12}$
10) $\frac{8}{9}$
11) $\frac{3}{2}$
12) $\frac{8}{11}$
13) $\frac{55}{7}$
14) $\frac{27}{25}$
15) 1
16) 3
17) $\frac{4}{3}$
18) $\frac{25}{63}$

Adding Mixed Numbers

1) 10
2) $5\frac{1}{2}$
3) $9\frac{3}{5}$
4) 4
5) $10\frac{2}{3}$
6) $4\frac{2}{3}$
7) $3\frac{8}{33}$
8) 4
9) $10\frac{4}{5}$
10) $7\frac{1}{5}$
11) $2\frac{1}{21}$
12) $3\frac{3}{4}$

Subtract Mixed Numbers

1) 1
2) $\frac{1}{4}$
3) $1\frac{2}{5}$
4) $\frac{2}{3}$
5) $\frac{2}{3}$
6) 2
7) $1\frac{19}{33}$
8) 1
9) $4\frac{2}{5}$
10) $6\frac{1}{5}$
11) $1\frac{8}{21}$
12) $\frac{3}{4}$

Multiplying Mixed Numbers

1) $2\frac{1}{12}$
2) $2\frac{2}{3}$
3) $5\frac{10}{21}$
4) $5\frac{31}{40}$
5) $7\frac{17}{25}$
6) $2\frac{2}{9}$
7) $4\frac{1}{16}$
8) $7\frac{12}{25}$
9) $11\frac{1}{3}$
10) $3\frac{9}{10}$
11) $1\frac{2}{3}$
12) $4\frac{2}{25}$

Dividing Mixed Numbers

1) $\frac{22}{25}$
2) $1\frac{19}{20}$
3) $\frac{19}{28}$
4) $\frac{1}{2}$
5) $1\frac{13}{20}$
6) $1\frac{9}{26}$
7) $2\frac{34}{63}$
8) $1\frac{11}{21}$
9) $2\frac{2}{15}$
10) $1\frac{34}{35}$
11) $3\frac{7}{10}$
12) 2

Comparing Decimals

1) 1.25 < 2.3
2) 0.5 > 0.23
3) 3.2 = 3.2
4) 4.58 < 45.8
5) 2.75 > 0.275
6) 5.2 > 5
7) 3.1 > 0.31
8) 6.33 > 0.733
9) 8 > 0.8
10) 4.56 > 0.456
11) 1.12 < 1.14
12) 2.77 < 2.78
13) 6.08 < 6.11
14) 1.11 > 0.211
15) 2.6 > 2.55
16) 1.24 < 1.25
17) 5.52 > 0.552
18) 0.33 > 0.033
19) 14.4 = 14.4
20) 0.05 < 0.50
21) 0.59 < 0.7
22) 0.5 > 0.05
23) 0.90 = 0.9
24) 0.27 < 0.4

Rounding Decimals

1) 0.2
2) 4.0
3) 5.6
4) 0.3
5) 6
6) 0.9
7) 8.2
8) 7
9) 1.63
10) 6.4
11) 2
12) 5
13) 5.9
14) 8.5
15) 81
16) 66
17) 70.8
18) 616
19) 16
20) 96
21) 2
22) 76
23) 116.5
24) 8.1

Adding and Subtracting Decimals

1) 2.96
2) 109.39
3) 94.84
4) 10.73
5) 147.34
6) 22.39
7) 3.5
8) 7.4
9) 6.1
10) 9.5
11) 3.5
12) 7.3

Multiplying and Dividing Decimals

1) 7.2
2) 76.23
3) 3.9
4) 86.33
5) 190.26
6) 22.77
7) 44.46
8) 9820
9) 23990
10) 2.5555…
11) 7.2631…
12) 2.6808…
13) 0.8024…
14) 0.14
15) 0.036
16) 0.424
17) 0.146
18) 0.0018

Converting Between Fractions, Decimals and Mixed Numbers

1) 0.9
2) 0.56
3) 0.75
4) 0.4
5) 0.333…
6) 0.8
7) 1.2
8) 1.6
9) 6.9
10) $\frac{3}{10}$
11) $4\frac{1}{2}$
12) $2\frac{1}{2}$
13) $2\frac{3}{10}$
14) $\frac{4}{5}$
15) $\frac{1}{4}$

16) $\frac{7}{50}$

17) $\frac{1}{5}$

18) $\frac{2}{25}$

19) $\frac{9}{20}$

20) $2\frac{3}{5}$

21) $5\frac{1}{5}$

Factoring Numbers

1) 1, 2, 4, 17, 34, 68
2) 1, 2, 4, 7, 8, 14, 28, 56
3) 1, 2, 3, 4, 6, 8, 12, 24
4) 1, 2, 4, 5, 8, 10, 20, 40
5) 1, 2, 43, 86
6) 1, 2, 3, 6, 13, 26, 39, 78
7) 1, 2, 5, 10, 25, 50
8) 1, 2, 7, 14, 49, 98
9) 1, 3, 5, 9, 15, 45
10) 1, 2, 13, 26
11) 1, 2, 3, 6, 9, 18, 27, 54
12) 1, 2, 4, 7, 14, 28
13) 1, 5, 11, 55
14) 1, 5, 17, 85
15) 1, 2, 3, 4, 6, 8, 12, 16, 24, 48
16) $2 \times 5 \times 5$
17) 5×5
18) 3×23
19) 3×7
20) $3 \times 3 \times 5$
21) $2 \times 2 \times 17$
22) 2×13
23) 2×43
24) 3×31

Greatest Common Factor

1) 10
2) 2
3) 5
4) 4
5) 1
6) 3
7) 3
8) 2
9) 2
10) 1
11) 2
12) 3
13) 4
14) 10
15) 1
16) 5
17) 2
18) 5
19) 12
20) 2
21) 10

Least Common Multiple

1) 28
2) 15
3) 80
4) 68
5) 24
6) 24
7) 18
8) 30
9) 152
10) 63
11) 551
12) 42
13) 150
14) 8
15) 150
16) 108
17) 72
18) 72
19) 780
20) 120
21) 90

Divisibility Rules

1) 16	**2** 3 **4** 5 6 7 **8** 9 10
2) 10	**2** 3 4 **5** 6 7 8 9 **10**
3) 15	2 **3** 4 **5** 6 7 8 9 10
4) 28	**2** 3 **4** 5 6 **7** 8 9 10
5) 36	**2** **3** **4** 5 **6** 7 8 **9** 10
6) 18	**2** **3** 4 5 **6** 7 8 **9** 10
7) 27	2 **3** 4 5 6 7 8 **9** 10
8) 70	**2** 3 4 **5** 6 **7** 8 9 **10**
9) 57	2 **3** 4 5 6 7 8 9 10
10) 102	**2** **3** 4 5 **6** 7 8 9 10
11) 144	**2** **3** **4** 5 **6** 7 **8** **9** 10
12) 75	2 **3** 4 **5** 6 7 8 9 10

Test Preparation Answers

1) Choice A is correct

2,500 out of 55,000 equals to $\frac{2500}{55000} = \frac{25}{550} = \frac{1}{22}$

2) The answer is 5^1.

Let x be the number of blank.

$360 = 2 \times 2 \times 2 \times 3 \times 3 \times x \Rightarrow x = \frac{360}{72} \Rightarrow x = 5$

3) Choice D is correct.

Compare each fraction, then we have:

$\frac{3}{10} < \frac{6}{13} < \frac{1}{2} < \frac{2}{3} < \frac{5}{7}$

4) Choice B is correct

To find the discount, multiply the 5 by $\frac{2}{3}$. Therefore we have $\frac{10}{3}$

After simplification now we have $3\frac{1}{3}$,that is between 3 and 4.

5) Choice D is correct

$4 \div \frac{1}{6} = 24$

6) Choice C is correct

Three times of 24,000 is 72,000. One sixth of them cancelled their tickets.

One sixth of 72,000 equals 12,000 (1/6 × 72,000 = 12,000).

60,000 (72000 – 12000 = 60000) fans are attending this week

7) Choice C is correct

Let x be the missed number. $180 = 4 \times 9 \times x \Rightarrow x=5$

Chapter 2: Integers and Absolute Value

Topics that you'll learn in this chapter:

- ✓ Adding and Subtracting Integers
- ✓ Multiplying and Dividing Integers
- ✓ Ordering Integers and Numbers
- ✓ Arrange, Order, and Comparing Integers
- ✓ Order of Operations
- ✓ Mixed Integer Computations
- ✓ Absolute Value
- ✓ Integers and Absolute Value
- ✓ Classifying Real Numbers Venn Diagram

Adding and Subtracting Integers

Helpful Hints	Integers: {… , −3, −2, −1, 0, 1, 2, 3, …} Includes: zero, counting numbers, and the negative of the counting numbers. – Add a positive integer by moving to the right on the number line. – Add a negative integer by moving to the left on the number line. – Subtract an integer by adding its opposite.	**Example:** 12 + 10 = 22 25 − 13 = 12 (−24) + 12 = −12 (−14) + (−12) = −26 14 − (−13) = 27

Find the sum.

1) (− 12) + (− 4)

2) 5 + (− 24)

3) (− 14) + 23

4) (− 8) + (39)

5) 43 + (−12)

6) (− 23) + (− 4) + 3

7) 4 + (− 12) + (− 10) + (− 25)

8) 19 + (− 15) + 25 + 11

9) (− 9) + (− 12) + (32 − 14)

10) 4 + (− 30) + (45 − 34)

Find the difference.

11) (− 14) − (− 9) − (18)

12) (− 9) − (− 25)

13) (− 12) − (8)

14) (28) − (− 4)

15) (34) − (2)

16) (55) − (− 5) + (− 4)

17) (9) − (2) − (− 5)

18) (2) − (4) − (− 15)

19) (23) − (4) − (− 34)

20) (− 45) − (− 87)

Multiplying and Dividing Integers

Helpful Hints

(negative) × (negative) = positive

(negative) ÷ (negative) = positive

(negative) × (positive) = negative

(negative) ÷ (positive) = negative

(positive) × (positive) = positive

Examples:

$3 \times 2 = 6$

$3 \times -3 = -9$

$-2 \times -2 = 4$

$10 \div 2 = 5$

$-4 \div 2 = -2$

$-12 \div -6 = 3$

Find each product.

1) $(-8) \times (-2)$
2) 3×6
3) $(-4) \times 5 \times (-6)$
4) $2 \times (-6) \times (-6)$
5) $11 \times (-12)$
6) $10 \times (-5)$
7) 8×8
8) $(-8) \times (-9)$
9) $6 \times (-5) \times 3$
10) $6 \times (-1) \times 2$

Find each quotient.

11) $18 \div 3$
12) $(-24) \div 4$
13) $(-63) \div (-9)$
14) $54 \div 9$
15) $20 \div (-2)$
16) $(-66) \div (-11)$
17) $64 \div 8$
18) $(-121) \div 11$
19) $72 \div 9$
20) $16 \div 4$

Ordering Integers and Numbers

Helpful Hints	To compare numbers, you can use number line! As you move from left to right on the number line, you find a bigger number!	**Example:** Order integers from least to greatest. (– 11, – 13, 7, – 2, 12) – 13 <– 11< – 2 < 7 <12

Order each set of integers from least to greatest.

1) – 15, – 19, 20, – 4, 1 ___, ___, ___, ___, ___, ___

2) 6, – 5, 4, – 3, 2 ___, ___, ___, ___, ___, ___

3) 15, – 42, 19, 0, – 22 ___, ___, ___, ___, ___, ___

4) 26, – 91, 0, – 13, 67, – 55 ___, ___, ___, ___, ___, ___

5) – 17, – 71, 90, – 25, – 54, – 39 ___, ___, ___, ___, ___, ___

6) 98, 5, 46, 19, 77, 24 ___, ___, ___, ___, ___, ___

Order each set of integers from greatest to least.

7) – 2, 5, – 3, 6, – 4 ___, ___, ___, ___, ___, ___

8) – 37, 7, – 17, 27, 47 ___, ___, ___, ___, ___, ___

9) 32, – 27, 19, – 17, 15 ___, ___, ___, ___, ___, ___

10) 68, 81, 21, – 18, 94, 72 ___, ___, ___, ___, ___, ___

Arrange, Order, and Comparing Integers

Helpful Hints	When using a number line, numbers increase as you move to the right.	**Examples:** 5 < 7, – 5 < – 2 – 18 < – 12

Arrange these integers in descending order.

1) 21, 71, – 18, – 10, 82 ___, ___, ___, ___, ___, ___

2) 15, 11, 20, 12, – 9, – 5 ___, ___, ___, ___, ___, ___

3) – 5, 20, 15, 9, –11 ___, ___, ___, ___, ___, ___

4) 19, 18, – 9, – 6, – 11 ___, ___, ___, ___, ___, ___

5) 56, – 34, – 12, – 5, 32 ___, ___, ___, ___, ___, ___

Compare. Use >, =, <

6) – 8 ____ 12

7) – 10 ____ –16

8) 43 ____ 34

9) 15 ____ –16

10) – 354 ____ –345

11) – 56 ____ – 58

12) 78 ____ 87

13) – 92 ____ – 102

14) – 12 ____ – 12

15) – 721 ____ – 821

Order of Operations

Helpful Hints	- Use "order of operations" rule when there are more than one math operation. - PEMDAS (parentheses / exponents / multiply / divide / add / subtract)	**Example:** $(12 + 4) \div (-4) = -4$

Evaluate each expression.

1) $(2 \times 2) + 5$

2) $24 - (3 \times 3)$

3) $(6 \times 4) + 8$

4) $25 - (4 \times 2)$

5) $(6 \times 5) + 3$

6) $64 - (2 \times 4)$

7) $25 + (1 \times 8)$

8) $(6 \times 7) + 7$

9) $48 \div (4 + 4)$

10) $(7 + 11) \div (-2)$

11) $9 + (2 \times 5) + 10$

12) $(5 + 8) \times \frac{3}{5} + 2$

13) $2 \times 7 - (\frac{10}{9-4})$

14) $(12 + 2 - 5) \times 7 - 1$

15) $(\frac{7}{5-1}) \times (2 + 6) \times 2$

16) $20 \div (4 - (10 - 8))$

17) $\frac{50}{4(5-4) - 3}$

18) $2 + (8 \times 2)$

Mixed Integer Computations

Helpful Hints

It worth remembering:

(negative) × (negative) = positive

(negative) ÷ (negative) = positive

(negative) × (positive) = negative

(negative) ÷ (positive) = negative

(positive) × (positive) = positive

Example:

$(-5) + 6 = 1$

$(-3) \times (-2) = 6$

$(9) \div (-3) = -3$

Compute.

1) $(-70) \div (-5)$

2) $(-14) \times 3$

3) $(-4) \times (-15)$

4) $(-65) \div 5$

5) $18 \times (-7)$

6) $(-12) \times (-2)$

7) $\frac{(-60)}{(-20)}$

8) $24 \div (-8)$

9) $22 \div (-11)$

10) $\frac{(-27)}{3}$

11) $4 \times (-4)$

12) $\frac{(-48)}{12}$

13) $(-14) \times (-2)$

14) $(-7) \times (7)$

15) $\frac{-30}{-6}$

16) $(-54) \div 6$

17) $(-60) \div (-5)$

18) $(-7) \times (-12)$

19) $(-14) \times 5$

20) $88 \div (-8)$

Absolute Value

Helpful Hints

Refers to the distance of a number from 0, the distances are positive. Therefore, absolute value of a number cannot be negative. $|-22| = 22$

Example:

$|12| \times |-2| = 24$

$$|x| = \begin{cases} x & for\ x \geq 0 \\ -x & for\ x < 0 \end{cases}$$

$|x| < n \Rightarrow -n < x < n$

$|x| > n \Rightarrow x < -n$ or $x > n$

Evaluate.

1) $|-4| + |-12| - 7$

2) $|-5| + |-13|$

3) $-18 + |-5 + 3| - 8$

4) $|27| \div |9|$

5) $|-9| \div |-1|$

6) $|200| \div |-100|$

7) $|55| \div |11|$

8) $|36| \div |-6|$

9) $|25| \times |-5|$

10) $|-3| \times |-8|$

11) $|12| \times |-5|$

12) $|11| \times |-6|$

13) $|-8| \times |4|$

14) $|-9| \times |-7|$

15) $|43 - 67 + 9| + |-11| - 1$

16) $|-45 + 78| + |23| - |45|$

17) $75 + |-11 - 30| - |2|$

18) $|-3 + 15| + |9 + 4| - 1$

Integers and Absolute Value

Helpful Hints	To find an absolute value of a number, just find it's distance from 0!	**Example:** $\lvert -6\rvert = 6$ $\lvert 6\rvert = 6$ $\lvert -12\rvert = 12$ $\lvert 12\rvert = 12$

Write absolute value of each number.

1) -4
2) -7
3) -8
4) 4
5) 5
6) -10
7) 1
8) 6
9) 8
10) -2
11) -1
12) 10
13) 3
14) 7
15) -5
16) -3
17) -9
18) 2
19) 4
20) -6
21) 9

Evaluate.

22) $\lvert -43\rvert - \lvert 12\rvert + 10$
23) $76 + \lvert -15 - 45\rvert - \lvert 3\rvert$
24) $30 + \lvert -62\rvert - 46$
25) $\lvert 32\rvert - \lvert -78\rvert + 90$
26) $\lvert -35 + 4\rvert + 6 - 4$
27) $\lvert -4\rvert + \lvert -11\rvert$
28) $\lvert -6 + 3 - 4\rvert + \lvert 7 + 7\rvert$
29) $\lvert -9\rvert + \lvert -19\rvert - 5$

Classifying Real Numbers Venn Diagram

Helpful Hints

Natural numbers (counting numbers): are the numbers that are used for counting. 1, 2, 3, …, 100, … are natural numbers.

Whole numbers are the natural numbers plus zero.

Integers include all whole numbers plus "negatives" of the natural numbers.

Rational numbers are numbers that can be written as a fraction. Both top and bottom numbers must be integers.

Irrational numbers are all numbers which cannot be written as fractions.

Real numbers include both the rational and irrational numbers.

Example:

0.25 = rational number and real number

Identify all of the subsets of real number system to which each number belongs.

Example:

0.1259 : Rational number

$\sqrt{2}$: Irrational number

3 : Natural number, whole number, Integer, rational number

1) 0
2) -5
3) -8.5
4) $\sqrt{4}$
5) -10
6) 18
7) 6
8) π
9) $1\frac{2}{7}$
10) -1
11) $\sqrt{5}$

Test Preparation

1) Which expression has a value of (-18)?

A. $8-(-4)+(-15)\times 2$

B. $12+(-3)\times(-2)$

C. $-6\times(-6)\times(-2)\div(-4)$

D. $(-2)\times(-7)+4$

E. $(-4)\times(-6)-6$

2) $4+8\times(3)-[8+8\times 5]\div 6=?$

Write your answer in the box below.

Answers of Worksheets – Chapter 2

Adding and Subtracting Integers

1) – 16
2) – 19
3) 9
4) 31
5) 31
6) – 24
7) – 43
8) 40
9) – 3
10) – 15
11) – 23
12) 16
13) – 20
14) 32
15) 32
16) 56
17) 12
18) 13
19) 53
20) 42

Multiplying and Dividing Integers

1) 16
2) 18
3) 120
4) 72
5) – 132
6) – 50
7) 64
8) 72
9) – 90
10) – 12
11) 6
12) – 6
13) 7
14) 6
15) – 10
16) 6
17) 8
18) – 11
19) 8
20) 4

Ordering Integers and Numbers

1) – 19, – 15, – 4, 1, 20
2) – 5, – 3, 2, 4, 6
3) – 42, – 22, 0, 15, 19
4) – 91, – 55, – 13, 0, 26, 67
5) – 71, – 54, – 39, – 25, – 17, 90
6) 5, 19, 24, 46, 77, 98
7) 6, 5, – 2, – 3, – 4
8) 47, 27, 7, – 17, – 37
9) 32, 19, 15, – 17, – 27
10) 94, 81, 72, 68, 21, – 18

Arrange and Order, Comparing Integers

1) 82, 71, 21, – 10, – 18
2) 20, 15, 12, 11, – 5, – 9
3) 20, 15, 9, – 5, –11
4) 19, 18, – 6, – 9, – 11
5) 56, 32, – 5, – 12, – 34
6) <
7) >
8) >
9) >
10) <
11) >
12) <
13) >
14) =
15) >

Order of Operations

1) 9
2) 15
3) 32
4) 17
5) 33
6) 56
7) 33
8) 49
9) 6
10) – 9
11) 29
12) 9.8
13) 12
14) 62
15) 28
16) 10
17) 50
18) 18

Mixed Integer Computations

1) 14
2) – 42
3) 60
4) – 13
5) – 126
6) 24
7) 3
8) – 3
9) – 2
10) – 9
11) – 16
12) – 4
13) 28
14) – 49
15) 5
16) – 9
17) 12
18) 84
19) – 70
20) – 11

Absolute Value

1) 9
2) 18
3) – 24
4) 3
5) 9
6) 2
7) 5
8) 6
9) 125
10) 24
11) 60
12) 66
13) 32
14) 63
15) 25
16) 11
17) 114
18) 24

Integers and Absolute Value

1) 4
2) 7
3) 8
4) 4
5) 5
6) 10
7) 1
8) 6
9) 8
10) 2
11) 1
12) 10
13) 3
14) 7
15) 5
16) 3
17) 9
18) 2
19) 4
20) 6
21) 9
22) 41
23) 133
24) 46
25) 44
26) 33
27) 15
28) 21
29) 23

Classifying Real Numbers Venn Diagram

1) 0: whole number, integer, rational number
2) – 5: integer, rational number
3) – 8.5: rational number
4) $\sqrt{4}$: natural number, whole number, integer, rational number
5) – 10: integer, rational number
6) 18 : natural number, whole number, integer, rational number
7) 6: natural number, whole number, integer, rational number
8) π: irrational number
9) $1\frac{2}{7}$: rational number
10) – 1: integer, rational number
11) $\sqrt{5}$: irrational number

Test Preparation Answers

1) Choice A is correct

Use PEMDAS (order of operation):

$8 - (-4) + (-15) \times 2 = 8 + 4 - 30 = -18$

2) The answer is 20.

Use PEMDAS (order of operation):

$4 + 8 \times (3) - [48] \div 6 = 4 + 24 - 8 = 20$

Chapter 3: Percent

Topics that you'll learn in this chapter:

- ✓ Percentage Calculations
- ✓ Converting Between Percent, Fractions, and Decimals
- ✓ Percent Problems
- ✓ Find What Percentage a Number Is of Another
- ✓ Find a Percentage of a Given Number
- ✓ Percent of Increase and Decrease

$$\frac{\text{is}}{\text{of}} = \frac{\%}{100} \quad \text{OR} \quad \frac{\text{Part}}{\text{whole}} = \frac{\%}{100}$$

Percentage Calculations

Helpful Hints

- Use the following formula to find part, whole, or percent:

$$\text{part} = \frac{\text{percent}}{100} \times \text{whole}$$

Example:

$$\frac{20}{100} \times 100 = 20$$

Calculate the percentages.

1) 50% of 25
2) 80% of 15
3) 30% of 34
4) 70% of 45
5) 10% of 0
6) 80% of 22
7) 65% of 8
8) 78% of 54
9) 50% of 80
10) 20% of 10
11) 40% of 40
12) 90% of 0
13) 20% of 70
14) 55% of 60
15) 80% of 10
16) 20% of 880
17) 70% of 100
18) 80% of 90

Solve.

19) 50 is what percentage of 75?
20) What percentage of 100 is 70
21) Find what percentage of 60 is 35.
22) 40 is what percentage of 80?

Converting Between Percent, Fractions, and Decimals

Helpful Hints		Examples:
	– To a percent: Move the decimal point 2 places to the right and add the % symbol.	
	– Divide by 100 to convert a number from percent to decimal.	30% = 0.3
		0.24 = 24%

Converting fractions to decimals.

1) $\frac{50}{100}$

4) $\frac{80}{100}$

7) $\frac{90}{100}$

2) $\frac{38}{100}$

5) $\frac{7}{100}$

8) $\frac{20}{100}$

3) $\frac{15}{100}$

6) $\frac{35}{100}$

9) $\frac{7}{100}$

Write each decimal as a percent.

10) 0.5

13) 0.524

16) 3.63

11) 0.9

14) 0.1

17) 0.008

12) 0.002

15) 0.03

18) 4.78

Percent Problems

Helpful Hints	Base = Part ÷ Percent Part = Percent × Base Percent = Part ÷ Base	**Example:** 2 is 10% of 20. 2 ÷ 0.10 = 20 2 = 0.10 × 20 0.10 = 2 ÷ 20

Solve each problem.

1) 51 is 340% of what?
2) 93% of what number is 97?
3) 27% of 142 is what number?
4) What percent of 125 is 29.3?
5) 60 is what percent of 126?
6) 67 is 67% of what?
7) 67 is 13% of what?
8) 41% of 78 is what?
9) 1 is what percent of 52.6?
10) What is 59% of 14 m?
11) What is 90% of 130 inches?
12) 16 inches is 35% of what?
13) 90% of 54.4 hours is what?
14) What percent of 33.5 is 21?

15) Liam scored 22 out of 30 marks in Algebra, 35 out of 40 marks in science and 89 out of 100 marks in mathematics. In which subject his percentage of marks in best?

16) Ella require 50% to pass. If she gets 280 marks and falls short by 20 marks, what were the maximum marks she could have got?

Find What Percentage a Number Is of Another

Helpful Hints	PERCENT: the number with the percent sign (%). PART: the number with the word "is". WHOLE: the number with the word "of". – Divide the Part by the Base. – Convert the answer to percent.	**Example:** 20 is what percent of 50? $20 \div 50 = 0.40 = 40\%$

Find the percentage of the numbers.

1) 5 is what percent of 90?
2) 15 is what percent of 75?
3) 20 is what percent of 400?
4) 18 is what percent of 90?
5) 3 is what percent of 15?
6) 8 is what percent of 80?
7) 11 is what percent of 55?
8) 9 is what percent of 90?
9) 2.5 is what percent of 10?
10) 5 is what percent of 25?
11) 60 is what percent of 20?
12) 12 is what percent of 48?
13) 14 is what percent of 28?
14) 8.2 is what percent of 32.8?
15) 1200 is what percent of 4,800?
16) 4,000 is what percent of 20,000?
17) 45 is what percent of 900?
18) 10 is what percent of 200?
19) 15 is what percent of 60?
20) 1.2 is what percent of 24?

Find a Percentage of a Given Number

Helpful Hints	- Use following formula to find part, whole, or percent: $\text{part} = \frac{\text{percent}}{100} \times \text{whole}$	**Example:** $\frac{50}{100} \times 50 = 25$

Find a Percentage of a Given Number.

1) 90% of 50

2) 40% of 50

3) 10% of 0

4) 80% of 80

5) 60% of 40

6) 50% of 60

7) 30% of 20

8) 35% of 10

9) 10% of 80

10) 10% of 60

11) 100% Of 50

12) 90% of 34

13) 80% of 42

14) 90% of 12

15) 20% of 56

16) 40% of 40

17) 40% of 6

18) 70% of 38

19) 30% of 3

20) 40% of 50

21) 100% of 8

Percent of Increase and Decrease

Helpful Hints	– To find the percentage increase: New Number – Original Number The result ÷ Original Number × 100 If your answer is a negative number, then this is a percentage decrease. To calculate percentage decrease: Original Number – New Number The result ÷ Original Number × 100	**Example:** From 84 miles to 24 miles = 71.43% decrease

Find each percent change to the nearest percent. Increase or decrease.

1) From 32 grams to 82 grams.
2) From 150 m to 45 m
3) From $438 to $443
4) From 256 ft to 140 ft
5) From 6469 ft to 7488 ft
6) From 36 inches to 90 inches
7) From 54 ft to 104 ft
8) From 84 miles to 24 miles
9) The population of a place in a particular year increased by 15%. Next year it decreased by 15%. Find the net increase or decrease percent in the initial population.

10) The salary of a doctor is increased by 40%. By what percent should the new salary be reduced in order to restore the original salary?

Test Preparation

1) Jason needs an 75% average in his writing class to pass. On his first 4 exams, he earned scores of 68%, 72%, 85%, and 90%. What is the minimum score Jason can earn on his fifth and final test to pass?

Write your answer in the box below.

2) A shirt costing $200 is discounted 15%. Which of the following expressions can be used to find the selling price of the shirt?

A. (200) (0.70)
B. (200) – 200 (0.30)
C. (200) (0.15) – (200) (0.15)
D. (200) (0.85)
E. (200) (0.15)

3) The price of a car was $20,000 in 2014 and $16,000 in 2015. What is the rate of depreciation of the price of the car per year?

A. 15 %
B. 20 %
C. 25 %
D. 30 %
E. 40 %

4) The price of a laptop is decreased by 10% to $360. What is its original price?

A. 320
B. 380
C. 400
D. 450
E. 600

5) A bank is offering 4.5% simple interest on a savings account. If you deposit $8,000, how much interest will you earn in five years?

A. $360
B. $720
C. $1,800
D. $3,600
E. 4,800

6) There are 60 boys and 90 girls in a class. Of these students, 18 boys and 12 girls write left–handed. What percentage of the students in this class write left–handed?

Write your answer in the box below.

7) 25 is What percent of 20?

A. 20 %
B. 25 %
C. 125 %
D. 150 %
E. 175%

8) A $40 shirt now selling for $28 is discounted by what percent?

A. 20 %
B. 30 %
C. 40 %
D. 60 %
E. 80 %

9) From last year, the price of gasoline has increased from $1.25 per gallon to $1.75 per gallon. The new price is what percent of the original price?

A. 72 %
B. 120 %
C. 140 %
D. 160 %
E. 180 %

10) Sophia purchased a sofa for $530.40. The sofa is regularly priced at $624. What was the percent discount Sophia received on the sofa?

A. 12%

B. 15%

C. 20%

D. 25%

E. 40 %

11) 55 students took an exam and 11 of them failed. What percent of the students passed the exam?

A. 20 %
B. 40 %
C. 60 %
D. 80 %
E. 85 %

12) The price of a sofa is decreased by 25% to $420. What was its original price?

A. $480

B. $520

C. $560

D. $600

E. $650

Answers of Worksheets – Chapter 3

Percentage Calculations

1) 12.5
2) 12
3) 10.2
4) 31.5
5) 0
6) 17.6
7) 5.2
8) 42.12
9) 40
10) 2
11) 16
12) 0
13) 14
14) 33
15) 8
16) 176
17) 70
18) 72
19) 67%
20) 70%
21) 58%
22) 50%

Converting Between Percent, Fractions, and Decimals

1) 0.5
2) 0.38
3) 0.15
4) 0.8
5) 0.07
6) 0.35
7) 0.9
8) 0.2
9) 0.07
10) 50%
11) 90%
12) 0.2%
13) 52.4%
14) 10%
15) 3%
16) 363%
17) 0.8%
18) 478%

Percent Problems

1) 15
2) 104.3
3) 38.34
4) 23.44%
5) 47.6%
6) 100
7) 515.4
8) 31.98
9) 1.9%
10) 8.3 m
11) 117 inches
12) 45.7inches
13) 49 hours
14) 62.7%
15) Mathematics
16) 600

Find What Percentage a Number Is of Another

1) 45 is what percent of 90? 50 %
2) 15 is what percent of 75? 20 %
3) 20 is what percent of 400? 5 %
4) 18 is what percent of 90? 20 %
5) 3 is what percent of 15? 20 %
6) 8 is what percent of 80? 10 %
7) 11 is what percent of 55? 20 %
8) 9 is what percent of 90? 10 %
9) 2.5 is what percent of 10? 25 %
10) 5 is what percent of 25? 20 %
11) 60 is what percent of 20? 300 %
12) 12 is what percent of 48? 25 %
13) 14 is what percent of 28? 50 %
14) 8.2 is what percent of 32.8? 25 %
15) 1200 is what percent of 4,800? 25 %
16) 4,000 is what percent of 20,000? 20 %
17) 45 is what percent of 900? 5 %
18) 10 is what percent of 200? 5 %
19) 15 is what percent of 60? 25 %
20) 1.2 is what percent of 24? 5 %

Find a Percentage of a Given Number

1) 45
2) 20
3) 0
4) 64
5) 24
6) 30
7) 6
8) 3.5
9) 8
10) 6
11) 50
12) 30.6
13) 33.6
14) 10.8
15) 11.2
16) 16
17) 2.4
18) 26.6
19) 0.9
20) 20
21) 8

Percent of Increase and Decrease

1) 156.25% increase
2) 70% decrease
3) 1.142% increase
4) 45.31% decrease
5) 15.75% increase
6) 150% increase
7) 92.6% increase
8) 71.43% decrease
9) 2.25% decrease
10) $28\frac{4}{7}\%$

Test Preparation Answers

1) The answer is 60.

Jason needs an 75% average to pass for five exams. Therefore, the sum of 5 exams must be at least 5 × 75 = 375

The sum of 4 exams is: 68 + 72 + 85 + 90 = 315.

The minimum score Jason can earn on his fifth and final test to pass is: 375 – 315 = 60

2) Choice D is correct

To find the discount, multiply the number by (100% – rate of discount).

Therefore, for the first discount we get: (200) (100% – 15%) = (200) (0.85) = 170

For the next 15 % discount: (200) (0.85) (0.85)

3) Choice B is correct

Let x be the rate of depreciation.

If the price of a car is decreased by \$4,000 to \$16,000, then:

$x\ \%\ of\ \$20.000 = \$4.000 \Rightarrow \frac{x}{100} 20.000 = 4.000 \Rightarrow x = 4.000 \div 200 = 20$

4) Choice C is correct

Let x be the original price.

If the price of a laptop is decreased by 10% to \$360, then:

$90\ \%\ of\ x = 360 \Rightarrow 0.90x = 360 \Rightarrow x = 360 \div 0.90 = 400$

5) Choice C is correct

Use simple interest formula: $I = prt$

(I = interest, p = principal, r = rate, t = time)

$$I = (8000)(0.045)(5) = 1800$$

6) The answer is 20%.

The whole number of students = 60 boys + 90 girls = 150

The number of students that write left-handed = 18 boys + 12 girls = 30

Let x be the percentage of the students that write left-handed.

$x\ \%\ of\ total\ students = \text{left} - \text{handed students} \Rightarrow x\ \%\ 150 = 30 \Rightarrow x = 3000 \div 150 = 20\%$

7) Choice C is correct

Use percent formula:

part = $\frac{\text{percent}}{100} \times$ whole

25 = $\frac{\text{percent}}{100} \times 20 \Rightarrow$ 25 = $\frac{\text{percent} \times 20}{100} \Rightarrow$ 25 = $\frac{\text{percent} \times 2}{10}$, multiply both sides by 10.

250 = $\text{percent} \times 2$, divide both sides by 2. 125 = percent

8) Choice B is correct

Use the formula for Percent of Change

$\frac{\text{New Value} - \text{Old Value}}{Old\ Value} \times 100\ \%$

$\frac{28-40}{40} \times 100\ \%$ = –30 % (negative sign here means that the new price is less than old price).

9) Choice C is correct

Use percent formula:

$\text{part} = \frac{\text{percent}}{100} \times \text{whole}$

$1.75 = \frac{\text{percent}}{100} \times 1.25 \Rightarrow 1.75 = \frac{\text{percent} \times 1.25}{100} \Rightarrow 175 = \text{percent} \times 1.25 \Rightarrow$

$\text{percent} = \frac{175}{1.25} = 140$

10) Choice B is correct.

The question is this: 530.40 is what percent of 624?

Use percent formula:

$\text{part} = \frac{\text{percent}}{100} \times \text{whole}$

$530.40 = \frac{\text{percent}}{100} \times 624 \Rightarrow 530.40 = \frac{\text{percent} \times 624}{100} \Rightarrow 53040 = \text{percent} \times 624 \Rightarrow$

$\text{percent} = \frac{53040}{624} = 85$

530.40 is 85 % of 624. Therefore, the discount is: 100% – 85% = 15%

11) Choice D is correct

The failing rate is 11 out of 55 = $\frac{11}{55}$

Change the fraction to percent:

$\frac{11}{55} \times 100\% = 20\%$

20 percent of students failed. Therefore, 80 percent of students passed the exam.

12) Choice C is correct.

Let x be the original price.

If the price of the sofa is decreased by 25% to \$420, then: $75\ \%\ of\ x = 420 \Rightarrow 0.75x = 420$ $\Rightarrow x = 420 \div 0.75 = 560$

Chapter 4: Algebraic Expressions

Topics that you'll learn in this chapter:

- ✓ Expressions and Variables
- ✓ Simplifying Variable Expressions
- ✓ Translate Phrases into an Algebraic Statement
- ✓ The Distributive Property
- ✓ Evaluating One Variable
- ✓ Evaluating Two Variables
- ✓ Combining like Terms

Expressions and Variables

Helpful Hints

A variable is a letter that represents unknown numbers. A variable can be used in the same manner as all other numbers:

Addition	2 + a	2 plus a
Subtraction	$y - 3$	y minus 3
Division	$\frac{4}{x}$	4 divided by x
Multiplication	5a	5 times a

Simplify each expression.

1) $x + 5x$, use $x = 5$

2) $8(-3x + 9) + 6$, use $x = 6$

3) $10x - 2x + 6 - 5$, use $x = 5$

4) $2x - 3x - 9$, use $x = 7$

5) $(-6)(-2x - 4y)$, use $x = 1$, $y = 3$

6) $8x + 2 + 4y$, use $x = 9$, $y = 2$

7) $(-6)(-8x - 9y)$, use $x = 5$, $y = 5$

8) $6x + 5y$, use $x = 7$, $y = 4$

Simplify each expression.

9) $5(-4 + 2x)$

10) $-3 - 5x - 6x + 9$

11) $6x - 3x - 8 + 10$

12) $(-8)(6x - 4) + 12$

13) $9(7x + 4) + 6x$

14) $(-9)(-5x + 2)$

Simplifying Variable Expressions

Helpful Hints	– Combine "like" terms. (values with same variable and same power) – Use distributive property if necessary. Distributive Property: $a(b+c) = ab + ac$	**Example:** $2x + 2(1 - 5x) =$ $2x + 2 - 10x = -8x + 2$

Simplify each expression.

1) $-2 - x^2 - 6x^2$

2) $3 + 10x^2 + 2$

3) $8x^2 + 6x + 7x^2$

4) $5x^2 - 12x^2 + 8x$

5) $2x^2 - 2x - x$

6) $(-6)(8x - 4)$

7) $4x + 6(2 - 5x)$

8) $10x + 8(10x - 6)$

9) $9(-2x - 6) - 5$

10) $3(x + 9)$

11) $7x + 3 - 3x$

12) $2.5x^2 \times (-8x)$

Simplify.

13) $-2(4 - 6x) - 3x$, $x = 1$

14) $2x + 8x$, $x = 2$

15) $9 - 2x + 5x + 2$, $x = 5$

16) $5(3x + 7)$, $x = 3$

17) $2(3 - 2x) - 4$, $x = 6$

18) $5x + 3x - 8$, $x = 3$

19) $x - 7x$, $x = 8$

20) $5(-2 - 9x)$, $x = 4$

Translate Phrases into an Algebraic Statement

Helpful Hints	**Translating key words and phrases into algebraic expressions:** **Addition:** plus, more than, the sum of, etc. **Subtraction:** minus, less than, decreased, etc. **Multiplication:** times, product, multiplied, etc. **Division:** quotient, divided, ratio, etc. **Example:** eight more than a number is 20 $8 + x = 20$

Write an algebraic expression for each phrase.

1) A number increased by forty–two.

2) The sum of fifteen and a number

3) The difference between fifty–six and a number.

4) The quotient of thirty and a number.

5) Twice a number decreased by 25.

6) Four times the sum of a number and – 12.

7) A number divided by – 20.

8) The quotient of 60 and the product of a number and – 5.

9) Ten subtracted from a number.

10) The difference of six and a number.

The Distributive Property

Helpful Hints	Distributive Property:	**Example:**
	$a(b+c)=ab+ac$	$3(4+3x)$ $=12+9x$

Use the distributive property to simply each expression.

1) $-(-2-5x)$
2) $(-6x+2)(-1)$
3) $(-5)(x-2)$
4) $-(7-3x)$
5) $8(8+2x)$
6) $2(12+2x)$
7) $(-6x+8)\,4$
8) $(3-6x)(-7)$
9) $(-12)(2x+1)$
10) $(8-2x)\,9$
11) $(-2x)(-1+9x)-4x(4+5x)$
12) $3(-5x-3)+4(6-3x)$
13) $(-2)(x+4)-(2+3x)$
14) $(-4)(3x-2)+6(x+1)$
15) $(-5)(4x-1)+4(x+2)$
16) $(-3)(x+4)-(2+3x)$

Evaluating One Variable

Helpful Hints		
	– To evaluate one variable expression, find the variable and substitute a number for that variable. – Perform the arithmetic operations.	**Example:** $4x + 8, x = 6$ $4(6) + 8 = 24 + 8 = 32$

Simplify each algebraic expression.

1) $9 - x, x = 3$

2) $x + 2, x = 5$

3) $3x + 7, x = 6$

4) $x + (-5), x = -2$

5) $3x + 6, x = 4$

6) $4x + 6, x = -1$

7) $10 + 2x - 6, x = 3$

8) $10 - 3x, x = 8$

9) $\frac{20}{x} - 3, x = 5$

10) $(-3) + \frac{x}{4} + 2x, x = 16$

11) $(-2) + \frac{x}{7}, x = 21$

12) $(-\frac{14}{x}) - 9 + 4x, x = 2$

13) $(-\frac{6}{x}) - 9 + 2x, x = 3$

14) $(-2) + \frac{x}{8}, x = 16$

Evaluating Two Variables

Helpful Hints	To evaluate an algebraic expression, substitute a number for each variable and perform the arithmetic operations.	**Example:** $2x + 4y - 3 + 2,$ $x = 5, y = 3$ $2(5) + 4(3) - 3 + 2$ $= 10$ $+ 12 - 3 + 2$ $= 21$

Simplify each algebraic expression.

1) $2x + 4y - 3 + 2,$

$x = 5,\ y = 3$

2) $(-\frac{12}{x}) + 1 + 5y,$

$x = 6,\ y = 8$

3) $(-4)(-2a - 2b),$

$a = 5,\ b = 3$

4) $10 + 3x + 7 - 2y,$

$x = 7,\ y = 6$

5) $9x + 2 - 4y,$

$x = 7,\ y = 5$

6) $6 + 3(-2x - 3y),$

$x = 9,\ y = 7$

7) $12x + y,$

$x = 4,\ y = 8$

8) $x \times 4 \div y,$

$x = 3,\ y = 2$

9) $2x + 14 + 4y,$

$x = 6,\ y = 8$

10) $4a - (5 - b),$

$a = 4,\ b = 6$

Combining like Terms

Helpful Hints

– Terms are separated by "+" and "–" signs.

– Like terms are terms with same variables and same powers.

– Be sure to use the "+" or "–" that is in front of the coefficient.

Example:

$22x + 6 + 2x =$

$24x + 6$

Simplify each expression.

1) 5 + 2x – 8

2) (– 2x + 6) 2

3) 7 + 3x + 6x – 4

4) (– 4) – (3)(5x + 8)

5) 9x – 7x – 5

6) x – 12x

7) 7 (3x + 6) + 2x

8) (– 11x) – 10x

9) 3x – 12 – 5x

10) 13 + 4x – 5

11) (– 22x) + 8x

12) 2 (4 + 3x) – 7x

13) (– 4x) – (6 – 14x)

14) 5 (6x – 1) + 12x

15) 22x + 6 + 2x

16) (– 13x) – 14x

17) (– 6x) – 9 + 15x

18) (– 6x) + 7x

19) (– 5x) + 12 + 7x

20) (– 3x) – 9 + 15x

21) 20x – 19x

Test Preparation

1) Which expression is equivalent to $38x$?

A. $(x \times 30) \times 8$
B. $(x \times 30) + 8$
C. $(x \times 30) + (x \times 8)$
D. $(x \times 3) + 8$
E. $(x \times 3) + 8x$

2) If $x = -8$, which equation is true?

A. $x(2x - 4) = 120$
B. $8(4 - x) = 96$
C. $2(4x + 6) = 79$
D. $6x + 2 = -50$
E. $12x + 2 = -100$

Answers of Worksheets – Chapter 4

Expressions and Variables

1) 30
2) –66
3) 41
4) –16
5) 84
6) 82
7) 510
8) 62
9) 10x – 20
10) 6 – 11x
11) 3x + 2
12) 44 – 48x
13) 69x + 36
14) 45x – 18

Simplifying Variable Expressions

1) $-7x^2 - 2$
2) $10x^2 + 5$
3) $15x^2 + 6x$
4) $-7x^2 + 8x$
5) $2x^2 - 3x$
6) – 48x +24
7) – 26x + 12
8) 90x – 48
9) – 18x – 59
10) 3x + 27
11) 4x + 3
12) $-20x^3$
13) 1
14) 20
15) 26
16) 80
17) – 22
18) 16
19) – 48
20) – 190

Translate Phrases into an Algebraic Statement

1) $x + 42$
2) $15 + x$
3) $56 - x$
4) $30/x$
5) 2x – 25
6) $4(x + (-12))$
7) $\frac{x}{-20}$
8) $\frac{60}{-5x}$
9) $x - 10$
10) $6 - x$

The Distributive Property

1) 5x + 2
2) 6x – 2
3) –5x + 10
4) 3x – 7
5) 16x + 64
6) 4x + 24
7) – 24x + 32
8) 42x – 21
9) – 24x – 12
10) – 18x + 72
11) $-38x^2 - 14x$
12) – 27x + 15
13) – 5x – 10
14) – 6x + 14
15) – 16x + 13
16) – 6x – 14

Evaluating One Variable

1) 6
2) 7
3) 25
4) −7
5) 18
6) 2
7) 10
8) −14
9) 1
10) 33
11) 1
12) −8
13) −5
14) 0
15) −176

Evaluating Two Variables

1) 21
2) 39
3) 64
4) 26
5) 45
6) −111
7) 56
8) 6
9) 58
10) 17

Combining like Terms

1) $2x - 3$
2) $-4x + 12$
3) $9x + 3$
4) $-15x - 28$
5) $2x - 5$
6) $-11x$
7) $23x + 42$
8) $-21x$
9) $-2x - 12$
10) $4x + 8$
11) $-14x$
12) $-x + 8$
13) $10x - 6$
14) $42x - 5$
15) $24x + 6$
16) $-27x$
17) $9x - 9$
18) x
19) $2x + 12$
20) $12x - 9$
21) x

Test Preparation Answers

1) Choice C is correct.

Let's can alternatives on by one.

A. ($30x$) × 8 = 240x

B. ($30x$) + 8 = $30x$ + 8

C. ($30x$) + ($8x$) = 38x

D. ($3x$) +8 = $3x$ +8

E. ($3x$) +8x = $11x$

2) Choice B is correct.

$x = -8$, then:

A. $(-8)(2(-8) - 4) = 120 \rightarrow 160 = 120$ Wrong!

B. $8\,(4 - (-8)) = 96 \rightarrow 96 = 96$ Correct!

C. $2\,(4(-8) + 6) = 79 \rightarrow -52 = 79$ Wrong!

D. $6(-8) + 2 = -50 \rightarrow -46 = -50$ Wrong!

E. $12x + 2 = -100 \rightarrow -94 = -100$ Wrong!

Chapter 5: Equations

Topics that you'll learn in this chapter:

- ✓ One–Step Equations
- ✓ One–Step Equation Word Problems
- ✓ Two–Step Equations
- ✓ Two–Step Equation Word Problems
- ✓ Multi–Step Equations

One–Step Equations

Helpful Hints	- The values of two expressions on both sides of an equation are equal. $ax + b = c$ - You only need to perform one Math operation in order to solve the equation.	**Example:** $-8x = 16$ $x = -2$

Solve each equation.

1) $x + 3 = 17$

2) $22 = (-8) + x$

3) $3x = (-30)$

4) $(-36) = (-6x)$

5) $(-6) = 4 + x$

6) $2 + x = (-2)$

7) $20x = (-220)$

8) $18 = x + 5$

9) $(-23) + x = (-19)$

10) $5x = (-45)$

11) $x - 12 = (-25)$

12) $x - 3 = (-12)$

13) $(-35) = x - 27$

14) $8 = 2x$

15) $(-6x) = 36$

16) $(-55) = (-5x)$

17) $x - 30 = 20$

18) $8x = 32$

19) $36 = (-4x)$

20) $4x = 68$

21) $30x = 300$

One–Step Equation Word Problems

Helpful Hints	– Define the variable. – Translate key words and phrases into math equation. – Isolate the variable and solve the equation.

Solve.

1) How many boxes of envelopes can you buy with $18 if one box costs $3?

2) After paying $6.25 for a salad, Ella has $45.56. How much money did she have before buying the salad?

3) How many packages of diapers can you buy with $50 if one package costs $5?

4) Last week James ran 20 miles more than Michael. James ran 56 miles. How many miles did Michael run?

5) Last Friday Jacob had $32.52. Over the weekend he received some money for cleaning the attic. He now has $44. How much money did he receive?

6) After paying $10.12 for a sandwich, Amelia has $35.50. How much money did she have before buying the sandwich?

Two–Step Equations

Helpful Hints

– You only need to perform two math operations (add, subtract, multiply, or divide) to solve the equation.

– Simplify using the inverse of addition or subtraction.

– Simplify further by using the inverse of multiplication or division.

Example:

$-2(x-1)=42$

$(x-1)=-21$

$x=-20$

Solve each equation.

1) $5(8+x)=20$

2) $(-7)(x-9)=42$

3) $(-12)(2x-3)=(-12)$

4) $6(1+x)=12$

5) $12(2x+4)=60$

6) $7(3x+2)=42$

7) $8(14+2x)=(-34)$

8) $(-15)(2x-4)=48$

9) $3(x+5)=12$

10) $\frac{3x-12}{6}=4$

11) $(-12)=\frac{x+15}{6}$

12) $110=(-5)(2x-6)$

13) $\frac{x}{8}-12=4$

14) $20=12+\frac{x}{4}$

15) $\frac{-24+x}{6}=(-12)$

16) $(-4)(5+2x)=(-100)$

17) $(-12x)+20=32$

18) $\frac{-2+6x}{4}=(-8)$

19) $\frac{x+6}{5}=(-5)$

20) $(-9)+\frac{x}{4}=(-15)$

Two–Step Equation Word Problems

Helpful Hints

– Translate the word problem into equations with variables.

– Solve the equations to find the solutions to the word problems.

Solve.

1) The sum of three consecutive even numbers is 48. What is the smallest of these numbers?

2) How old am I if 400 reduced by 2 times my age is 244?

3) For a field trip, 4 students rode in cars and the rest filled nine buses. How many students were in each bus if 472 students were on the trip?

4) The sum of three consecutive numbers is 72. What is the smallest of these numbers?

5) 331 students went on a field trip. Six buses were filled, and 7 students traveled in cars. How many students were in each bus?

6) You bought a magazine for $5 and four erasers. You spent a total of $25. How much did each eraser cost?

Multi–Step Equations

Helpful Hints	– Combine "like" terms on one side. – Bring variables to one side by adding or subtracting. – Simplify using the inverse of addition or subtraction. – Simplify further by using the inverse of multiplication or division.	**Example:** $3x + 15 = -2x + 5$ Add 2x both sides $5x + 15 = +5$ Subtract 15 both sides $5x = -10$ Divide by 5 both sides $x = -2$

Solve each equation.

1) $-(2 - 2x) = 10$

2) $-12 = -(2x + 8)$

3) $3x + 15 = (-2x) + 5$

4) $-28 = (-2x) - 12x$

5) $2(1 + 2x) + 2x = -118$

6) $3x - 18 = 22 + x - 3 + x$

7) $12 - 2x = (-32) - x + x$

8) $7 - 3x - 3x = 3 - 3x$

9) $6 + 10x + 3x = (-30) + 4x$

10) $(-3x) - 8(-1 + 5x) = 352$

11) $24 = (-4x) - 8 + 8$

12) $9 = 2x - 7 + 6x$

13) $6(1 + 6x) = 294$

14) $-10 = (-4x) - 6x$

15) $4x - 2 = (-7) + 5x$

16) $5x - 14 = 8x + 4$

17) $40 = -(4x - 8)$

18) $(-18) - 6x = 6(1 + 3x)$

19) $x - 5 = -2(6 + 3x)$

20) $6 = 1 - 2x + 5$

Test Preparation

1) What is the value of x in the following equation?

$$x + \frac{1}{6} = \frac{1}{3}$$

A. 6

B. $\frac{1}{2}$

C. $\frac{1}{6}$

D. $\frac{1}{4}$

E. $\frac{1}{8}$

2) The area of a rectangle is x square feet and its length is 9 feet. Which equation represents y, the width of the rectangle in feet?

A. $y = \frac{x}{9}$

B. $y = \frac{9}{x}$

C. $y = 9x$

D. $y = 9 + x$

E. $y = 9 - x$

3) An angle is equal to one fifth of its supplement. What is the measure of that angle?

A. 20
B. 30
C. 45
D. 60
E. 80

4) In five successive hours, a car travels 40 km, 45 km, 50 km, 35 km and 55 km. In the next five hours, it travels with an average speed of 50 km per hour. Find the total distance the car traveled in 10 hours.

A. 425 km
B. 450 km
C. 475 km
D. 500 km
E. 550 km

5) In a triangle ABC the measure of angle ACB is 75° and the measure of angle CAB is 45°. What is the measure of angle ABC?

Write your answer in the box below.

6) If 40 % of a number is 4, what is the number?

A. 4
B. 8
C. 10
D. 12
E. 16

7) Jason is 9 miles ahead of Joe running at 5.5 miles per hour and Joe is running at the speed of 7 miles per hour. How long does it take Joe to catch Jason?

A. 3 hours
B. 4 hours
C. 6 hours
D. 8 hours
E. 10 hours

8) What is the equivalent temperature of 104°F in Celsius?

$$C = \frac{5}{9}(F - 32)$$

A. 32
B. 40
C. 48
D. 52
E. 56

9) If 150 % of a number is 75, then what is the 90 % of that number?

A. 45

B. 50

C. 70

D. 85

E. 90

Answers of Worksheets – Chapter 5

One–Step Equations

1) 14
2) 30
3) -10
4) 6
5) -10
6) -4
7) -11
8) 13
9) 4
10) -9
11) -13
12) -9
13) -8
14) 4
15) -6
16) 11
17) 50
18) 4
19) -9
20) 17
21) 10

One–Step Equation Word Problems

1) 6
2) $51.81
3) 10
4) 36
5) 11.48
6) 45.62

Two–Step Equations

1) -4
2) 3
3) 2
4) 1
5) 0.5
6) $\frac{4}{3}$
7) $-\frac{73}{8}$
8) $\frac{2}{5}$
9) -1
10) 12
11) -87
12) -8
13) 128
14) 32
15) -48
16) 10
17) -1
18) -5
19) -31
20) -24

Two–Step Equation Word Problems

1) 14
2) 78
3) 52
4) 23
5) 54
6) $4

Multi–Step Equations

1) 6
2) 2
3) -2
4) 2
5) -20
6) 37
7) 22
8) $\frac{4}{3}$
9) -4
10) -8
11) -6
12) 2
13) 8
14) 1
15) 5
16) -6
17) -8
18) -1
19) -1
20) 0

Test Preparation Answers

1) The answer is C.

Subtract the numerators and find x.

$$x = \frac{1}{3} - \frac{1}{6} \Rightarrow x = \frac{2-1}{6} \Rightarrow x = \frac{1}{6}$$

2) Choice A is correct

To find the area of rectangle = width multiply length

Therefore; $x = 9 \times y$

Then find y : $y = \frac{x}{9}$

3) Choice B is correct

The sum of supplement angles is 180. Let x be that angle. Therefore,

$x + 5x = 180$

$6x = 180$, divide both sides by 6: $x = 30$

4) Choice C is correct

Add the first 5 numbers. 40 + 45 + 50 + 35 + 55 = 225

To find the distance traveled in the next 5 hours, multiply the average by number of hours.

Distance = Average × Rate = 50 × 5 = 250

Add both numbers.

250 + 225 = 475

5) The answer is 60.

The whole angles in every triangle are:180°and Let x be the number of new angle so:

$180 = 75 + 45 + x \Rightarrow x = 60°$

6) Choice C is correct

Let x be the number. Write the equation and solve for x.

$40\%\ of\ x = 4 \Rightarrow 0.40\,x = 4 \Rightarrow x = 4\ \div 0.40 = 10$

7) Choice C is correct

The distance between Jason and Joe is 9 miles. Jason running at 5.5 miles per hour and Joe is running at the speed of 7 miles per hour. Therefore, every hour the distance is 1.5 miles less. 9 ÷ 1.5 = 6

8) Choice B is correct

Plug in 104 for F and then solve for C.

$$C = \frac{5}{9}(F - 32) \Rightarrow C = \frac{5}{9}(104 - 32) \Rightarrow C = \frac{5}{9}(72) = 40$$

9) Choice A is correct

First, find the number.

Let x be the number. Write the equation and solve for x.

150 % of a number is 75, then:

$1.5 \times x = 75 \Rightarrow x = 75 \div 1.5 = 50$

90 % of 50 is:

$0.9 \times 50 = 45$

Chapter 6: Proportions and Ratios

Topics that you'll learn in this chapter:

- ✓ Writing Ratios
- ✓ Simplifying Ratios
- ✓ Proportional Ratios
- ✓ Create a Proportion
- ✓ Similar Figures
- ✓ Similar Figure Word Problems
- ✓ Ratio and Rates Word Problems

Writing Ratios

Helpful Hints	– A ratio is a comparison of two numbers. Ratio can be written as a division.	**Example:** $3:5$, or $\frac{3}{5}$

Express each ratio as a rate and unite rate.

1) 120 miles on 4 gallons of gas.

2) 24 dollars for 6 books.

3) 200 miles on 14 gallons of gas

4) 24 inches of snow in 8 hours

Express each ratio as a fraction in the simplest form.

5) 3 feet out of 30 feet

6) 18 cakes out of 42 cakes

7) 16 dimes t0 24 dimes

8) 12 dimes out of 48 coins

9) 14 cups to 84 cups

10) 45 gallons to 65 gallons

11) 10 miles out of 40 miles

12) 22 blue cars out of 55 cars

13) 32 pennies to 300 pennies

14) 24 beetles out of 86 insects

Proportional Ratios

Helpful Hints

– A proportion means that two ratios are equal. It can be written in two ways:

$\frac{a}{b} = \frac{c}{d}$, a : b = c : d

Example:

$\frac{9}{3} = \frac{6}{d}$

$d = 2$

Solve each proportion.

1) $\frac{3}{6} = \frac{8}{d}$

2) $\frac{k}{5} = \frac{12}{15}$

3) $\frac{30}{5} = \frac{12}{x}$

4) $\frac{x}{2} = \frac{1}{8}$

5) $\frac{d}{3} = \frac{2}{6}$

6) $\frac{27}{7} = \frac{30}{x}$

7) $\frac{8}{5} = \frac{k}{15}$

8) $\frac{60}{20} = \frac{3}{d}$

9) $\frac{x}{3} = \frac{12}{18}$

10) $\frac{25}{5} = \frac{x}{8}$

11) $\frac{12}{x} = \frac{4}{2}$

12) $\frac{x}{4} = \frac{18}{2}$

13) $\frac{80}{10} = \frac{k}{10}$

14) $\frac{12}{6} = \frac{6}{d}$

15) $\frac{x}{4} = \frac{30}{5}$

16) $\frac{9}{5} = \frac{k}{5}$

17) $\frac{45}{15} = \frac{15}{d}$

18) $\frac{60}{x} = \frac{10}{3}$

19) $\frac{d}{3} = \frac{14}{6}$

20) $\frac{k}{4} = \frac{4}{2}$

21) $\frac{4}{2} = \frac{x}{7}$

Simplifying Ratios

Helpful Hints	– You can calculate equivalent ratios by multiplying or dividing both sides of the ratio by the same number.	**Examples:** 3 : 6 = 1 : 2 4 : 9 = 8 : 18

Reduce each ratio.

1) 21 : 49
2) 20 : 40
3) 10: 50
4) 14: 18
5) 45: 27
6) 49: 21
7) 100: 10
8) 12 : 8
9) 35 : 45
10) 8: 20
11) 25: 35
12) 21 : 27
13) 52 : 82
14) 12: 36
15) 24 : 3
16) 15: 30
17) 3 : 36
18) 8 : 16
19) 6 : 100
20) 2 : 20
21) 10: 60
22) 14: 63
23) 68: 80
24) 8: 80

Create a Proportion

Helpful Hints

– A proportion contains 2 equal fractions! A proportion simply means that two fractions are equal.

Example:

2, 4, 8, 16

$\frac{2}{4} = \frac{8}{16}$

Create proportion from the given set of numbers.

1) 1, 6, 2, 3

2) 12, 144, 1, 12

3) 16, 4, 8, 2

4) 9, 5, 27, 15

5) 7, 10, 60, 42

6) 8, 7, 24, 21

7) 10, 5, 8, 4

8) 3, 12, 8, 2

9) 2, 2, 1, 4

10) 3, 6, 7, 14

11) 2, 6, 5, 15

12) 7, 2, 14, 4

Similar Figures

Helpful Hints	– Two or more figures are similar if the corresponding angles are equal, and the corresponding sides are in proportion.	**Example:** 3–4–5 triangle is similar to a 6–8–10 triangle

Each pair of figures is similar. Find the missing side.

1)

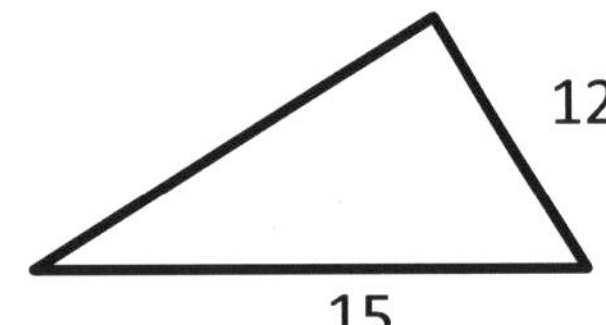

4

x

2)

5x

8

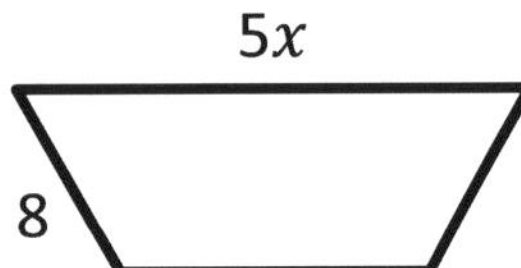

60

32

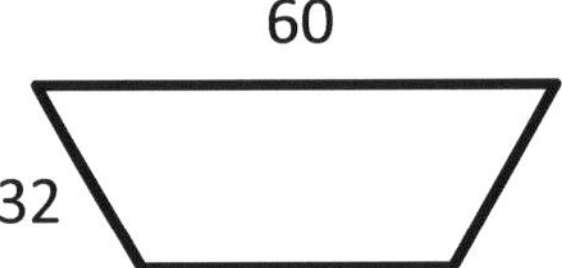

3)

40

56

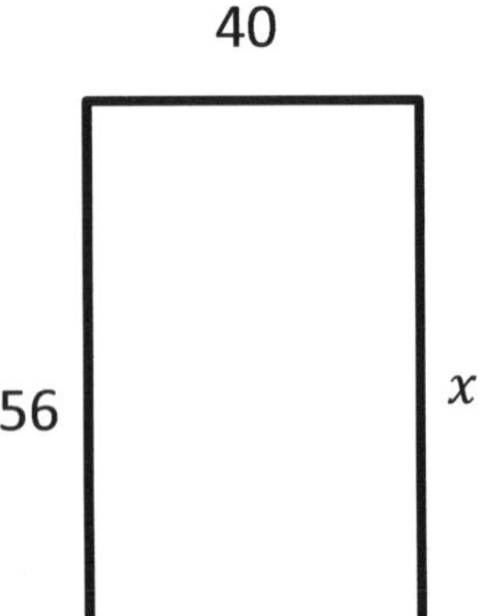

x

7

5

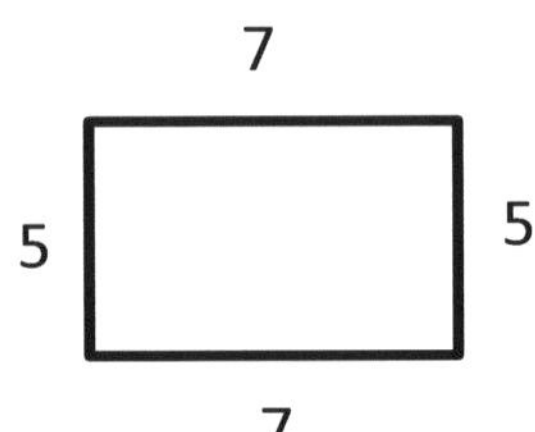

5

7

Similar Figure Word Problems

Helpful Hints

To solve a similarity word problem, create a proportion and use cross multiplication method!

Example:

$\frac{x}{4} = \frac{8}{16}$

$16x = 4 \times 8$

$x = 2$

Answer each question and round your answer to the nearest whole number.

1) If a 42.9 ft tall flagpole casts a 253.1 ft long shadow, then how long is the shadow that a 6.2 ft tall woman casts?

2) A model igloo has a scale of 1 in : 2 ft. If the real igloo is 10 ft wide, then how wide is the model igloo?

3) If a 18 ft tall tree casts a 9 ft long shadow, then how tall is an adult giraffe that casts a 7 ft shadow?

4) Find the distance between San Joe and Mount Pleasant if they are 2 cm apart on a map with a scale of 1 cm : 9 km.

5) A telephone booth that is 8 ft tall casts a shadow that is 4 ft long. Find the height of a lawn ornament that casts a 2 ft shadow.

Ratio and Rates Word Problems

Helpful Hints

To solve a ratio or a rate word problem, create a proportion and use cross multiplication method!

Example:

$\frac{x}{4} = \frac{8}{16}$

$16x = 4 \times 8$

$x = 2$

Solve.

1) In a party, 10 soft drinks are required for every 12 guests. If there are 252 guests, how many soft drink is required?

2) In Jack's class, 18 of the students are tall and 10 are short. In Michael's class 54 students are tall and 30 students are short. Which class has a higher ratio of tall to short students?

3) Are these ratios equivalent?
   12 cards to 72 animals 11 marbles to 66 marbles

4) The price of 3 apples at the Quick Market is \$1.44. The price of 5 of the same apples at Walmart is \$2.50. Which place is the better buy?

5) The bakers at a Bakery can make 160 bagels in 4 hours. How many bagels can they bake in 16 hours? What is that rate per hour?

6) You can buy 5 cans of green beans at a supermarket for \$3.40. How much does it cost to buy 35 cans of green beans?

Test Preparation

1) The ratio of boys and girls in a class is 4:7. If there are 44 students in the class, how many more boys should be enrolled to make the ratio 1:1?

   A. 8

   B. 10

   C. 12

   D. 14

   E. 16

2) In a party, 6 soft drinks are required for every 9 guests. If there are 171 guests, how many soft drink is required?

   A. 9
   B. 27
   C. 114
   D. 171
   E. 189

3) Peter traveled 150 km in 6 hours and Jason traveled 180 km in 4 hours. What is the ratio of the average speed of Peter to average speed of Jason?

   A. 3 : 2
   B. 2 : 3
   C. 5 : 9
   D. 5 : 6
   E. 5 : 8

4) How long does a 420–miles trip take moving at 50 miles per hour (mph)?

   A. 4 hours
   B. 6 hours and 24 minutes
   C. 8 hours and 24 minutes
   D. 8 hours and 30 minutes
   E. 9 hours

5) The ratio of boys to girls in a school is 3:4. If there are 350 students in a school, how many boys are in the school.

   Write your answer in the box below.

Answers of Worksheets – Chapter 6

Writing Ratios

1) $\frac{120\ miles}{4\ gallons}$, 30 miles per gallon

2) $\frac{24\ dollars}{6\ books}$, 4.00 dollars per book

3) $\frac{200\ miles}{14\ gallons}$, 14.29 miles per gallon

4) $\frac{24''\ of\ snow}{8\ hours}$, 3 inches of snow per hour

5) $\frac{1}{10}$
6) $\frac{3}{7}$
7) $\frac{2}{3}$
8) $\frac{1}{4}$
9) $\frac{1}{6}$
10) $\frac{9}{13}$
11) $\frac{1}{4}$
12) $\frac{2}{5}$
13) $\frac{8}{75}$
14) $\frac{12}{43}$

Simplifying Ratios

1) 3 : 7
2) 1 : 2
3) 1 : 5
4) 7 : 9
5) 5 : 3
6) 7 : 3
7) 10 : 1
8) 3 : 2
9) 7 : 9
10) 2 : 5
11) 5 : 7
12) 7 : 9
13) 26 : 41
14) 1 : 3
15) 8 : 1
16) 1 : 2
17) 1 : 12
18) 1 : 2
19) 3 : 50
20) 1 : 10
21) 1: 6
22) 2 : 9
23) 17 : 20
24) 1 : 10

Proportional Ratios

1) 16
2) 4
3) 2
4) 0.25
5) 1
6) 7.78
7) 24
8) 1
9) 2
10) 40
11) 6
12) 36

13) 80
14) 3
15) 24
16) 9
17) 5
18) 18
19) 7
20) 8
21) 14

Create a Proportion

1) 1 : 3 = 2 : 6
2) 12 : 144 = 1 : 12
3) 2 : 4 = 8 : 16
4) 5 : 15 = 9 : 27
5) 7 : 42, 10 : 60
6) 7 : 21 = 8 : 24
7) 8 : 10 = 4 : 5
8) 2 : 3 = 8 : 12
9) 4 : 2 = 2 : 1
10) 7 : 3 = 14 : 6
11) 5 : 2 = 15 : 6
12) 7 : 2 = 14 : 4

Similar Figures

1) 5
2) 3
3) 56

Similar Figure Word Problems

1) 36.6 ft
2) 5 in
3) 14 ft
4) 18 km
5) 4 ft

Ratio and Rates Word Problems

1) 210
2) The ratio for both class is equal to 9 to 5.
3) Yes! Both ratios are 1 to 6
4) The price at the Quick Market is a better buy.
5) 640, the rate is 40 per hour.
6) $23.80

Test Preparation Answers

1) Choice C is correct

The ratio of boy to girls is 4:7. Therefore, there are 4 boys out of 11 students. To find the answer, first divide the total number of students by 11, then multiply the result by 4.

$44 \div 11 = 4 \Rightarrow 4 \times 4 = 16$

There are 16 boys and 28 (44 – 16) girls. So, 12 more boys should be enrolled to make the ratio 1:1

2) Choice C is correct

Let x be the number of soft drinks for 171 guests. It's needed to have a proportional ratio to find x.

$$\frac{6 \text{ soft drinks}}{9 \text{ guests}} = \frac{x}{171 \text{ guests}}$$

$$x = \frac{171 \times 6}{9} \Rightarrow x = 114$$

3) Choice C is correct

Speed= $\frac{\textit{The amount of Km}}{\text{The amount of hours}}$

Peter's speed= $\frac{150}{6}$ =25

Jason's speed = $\frac{180}{4}$ =45

$\frac{\textit{The average speed of peter}}{\textit{The average speed of Jason}} = \frac{25}{45}$ and after simplification we have: $\frac{5}{9}$

4) Choice C is correct

Use distance formula:

Distance = Rate × time ⇒ 420 = 50 × T, divide both sides by 50. 420 / 50 = T ⇒ T = 8.4 hours.

Change hours to minutes for the decimal part. 0.4 hours = 0.4 × 60 = 24 minutes.

5) The answer is 150.

The ratio of boy to girls is 3:4. Therefore, there are 3 boys out of 7 students. To find the answer, first divide the total number of students by 7, then multiply the result by 3.

$350 \div 7 = 50 \Rightarrow 50 \times 3 = 150$

Chapter 7: Inequalities

Topics that you'll learn in this chapter:

- ✓ Graphing Single– Variable Inequalities
- ✓ One– Step Inequalities
- ✓ Two– Step Inequalities
- ✓ Multi– Step Inequalities

Graphing Single–Variable Inequalities

Helpful Hints	– Isolate the variable. – Find the value of the inequality on the number line. – For less than or greater than draw open circle on the value of the variable. – If there is an equal sign too, then use filled circle. – Draw a line to the right direction.

Draw a graph for each inequality.

1) $-2 > x$

-10 -9 -8 -7 -6 -5 -4 -3 -2 -1 0 1 2 3 4 5 6 7 8 9 10

2) $5 \leq -x$

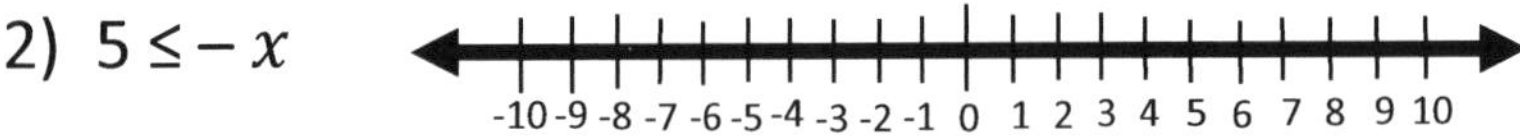

3) $x > 7$

4) $-x > 1.5$

One–Step Inequalities

Helpful Hints	– Isolate the variable. – For dividing both sides by negative numbers, flip the direction of the inequality sign.	**Example:** $x + 4 \geq 11$ $x \geq 7$

Solve each inequality and graph it.

1) $x + 9 \geq 11$

2) $x - 4 \leq 2$

3) $6x \geq 36$

4) $7 + x < 16$

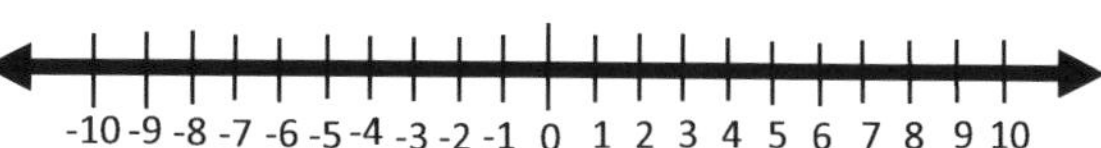

5) $x + 8 \leq 1$

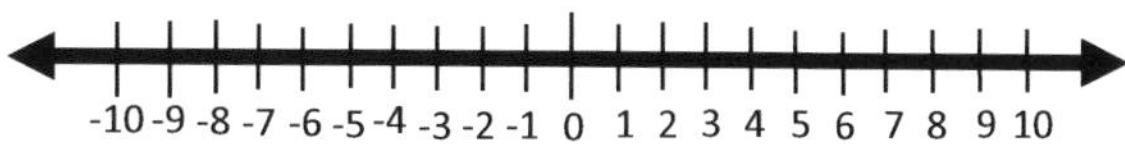

6) $3x > 12$

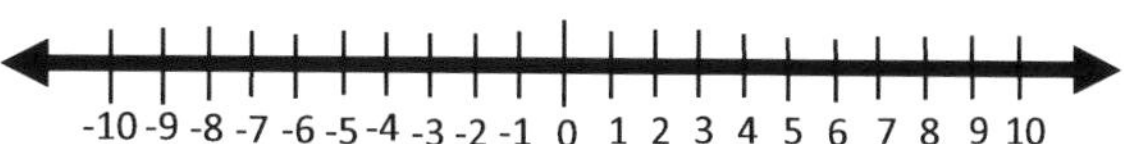

7) $3x < 24$

Two–Step Inequalities

Helpful Hints	– Isolate the variable. – For dividing both sides by negative numbers, flip the direction of the of the inequality sign. – Simplify using the inverse of addition or subtraction. – Simplify further by using the inverse of multiplication or division.	**Example:** $2x + 9 \geq 11$ $2x \geq 2$ $x \geq 1$

Solve each inequality and graph it.

1) $3x - 4 \leq 5$

2) $2x - 2 \leq 6$

3) $4x - 4 \leq 8$

4) $3x + 6 \geq 12$

5) $6x - 5 \geq 19$

6) $2x - 4 \leq 6$

7) $8x - 4 \leq 4$

8) $6x + 4 \leq 10$

9) $5x + 4 \leq 9$

10) $7x - 4 \leq 3$

11) $4x - 19 < 19$

12) $2x - 3 < 21$

13) $7 + 4x \geq 19$

14) $9 + 4x < 21$

15) $3 + 2x \geq 19$

16) $6 + 4x < 22$

Multi–Step Inequalities

Helpful Hints

– Isolate the variable.

– Simplify using the inverse of addition or subtraction.

– Simplify further by using the inverse of multiplication or division.

Example:

$\frac{7x+1}{3} \geq 5$

$7x + 1 \geq 15$

$7x \geq 14$

$x \geq 7$

Solve each inequality.

1) $\frac{9x}{7} - 7 < 2$

2) $\frac{4x+8}{2} \leq 12$

3) $\frac{3x-8}{7} > 1$

4) $-3\,(x - 7) > 21$

5) $4 + \frac{x}{3} < 7$

6) $\frac{2x+6}{4} \leq 10$

Test Preparation

1) A football team had $20,000 to spend on supplies. The team spent $14,000 on new balls. New sport shoes cost $120 each. Which of the following inequalities represent how many new shoes the team can purchase.

A. $120x + 14{,}000 \leq 20{,}000$

B. $120x + 14{,}000 \geq 20{,}000$

C. $14{,}000x + 120 \leq 20{,}000$

D. $14{,}000x + 120 \geq 20{,}000$

E. $14{,}000x - 120 \geq 20{,}000$

2) In 1999, the average worker's income increased $2,000 per year starting from $24,000 annual salary. Which equation represents income greater than average? (I = income, x = number of years after 1999)

A. I > 2000 x + 24000

B. I > – 2000 x + 24000

C. I < –2000 x + 24000

D. I < 2000 x – 24000

E. I < 2000 x + 24000

3) Which of the following graphs represents the compound inequality $-2 \leq 2x - 4 < 8$?

A. 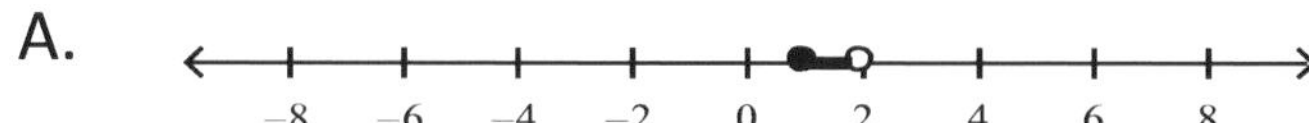

B. 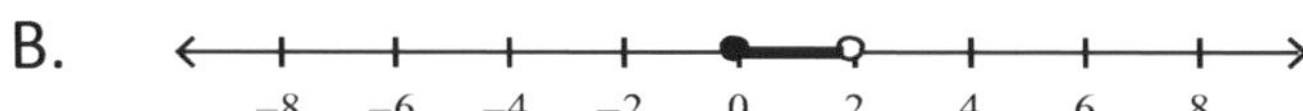

C. 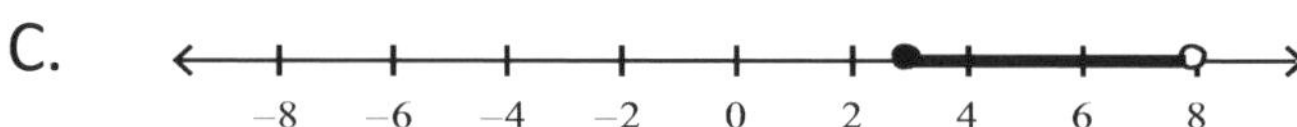

D. 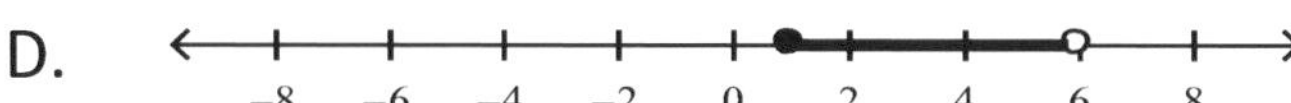

Answers of Worksheets – Chapter 7

Graphing Single–Variable Inequalities

1) $-2 > x$

2) $x \leq -5$

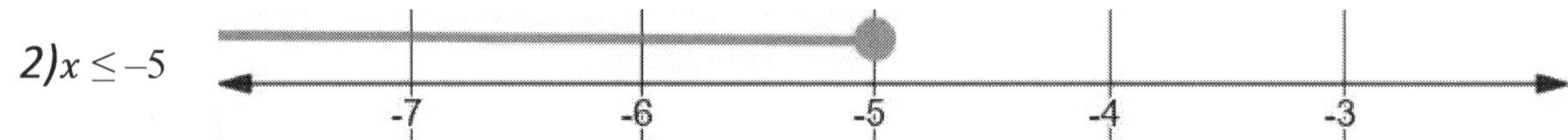

3) $x > 7$

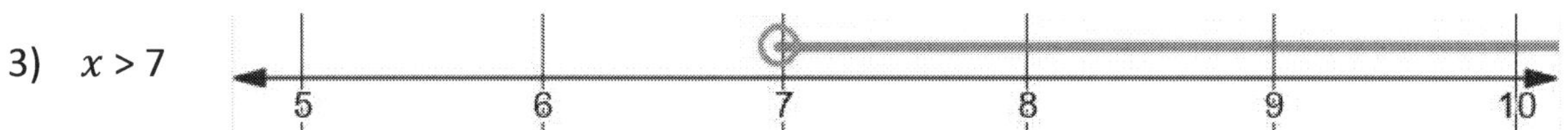

4) $-1.5 > x$

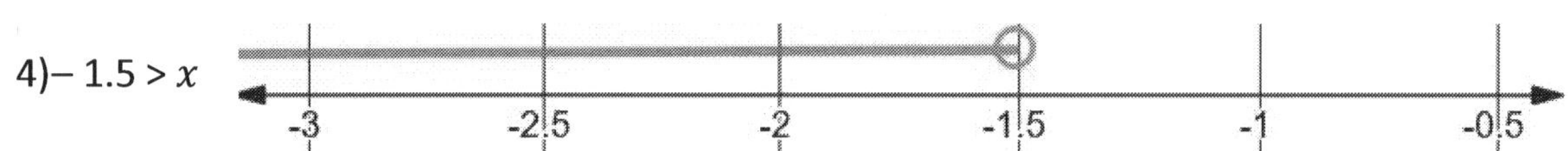

One–Step Inequalities

1)

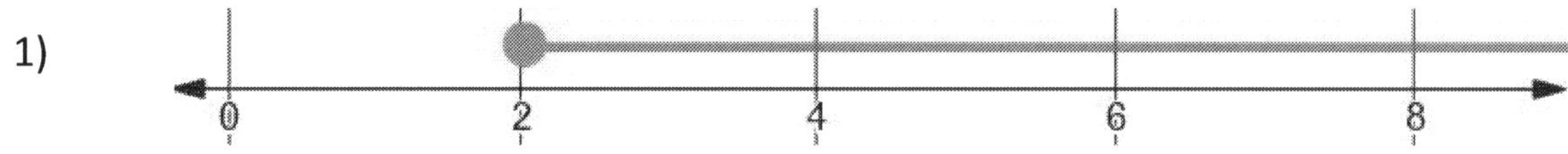

2)

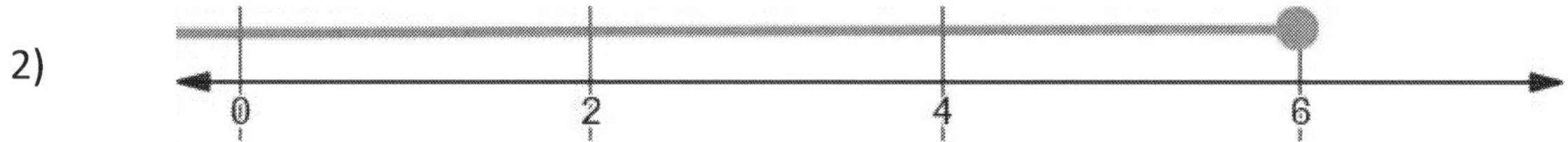

3)

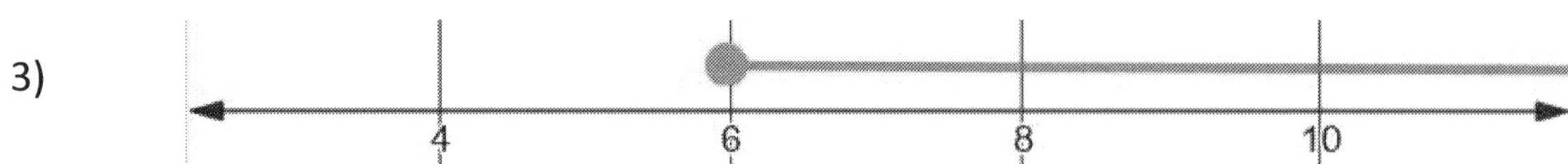

4) 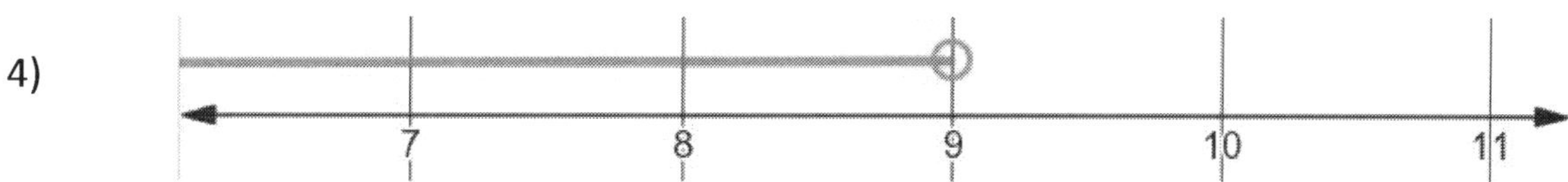

5) 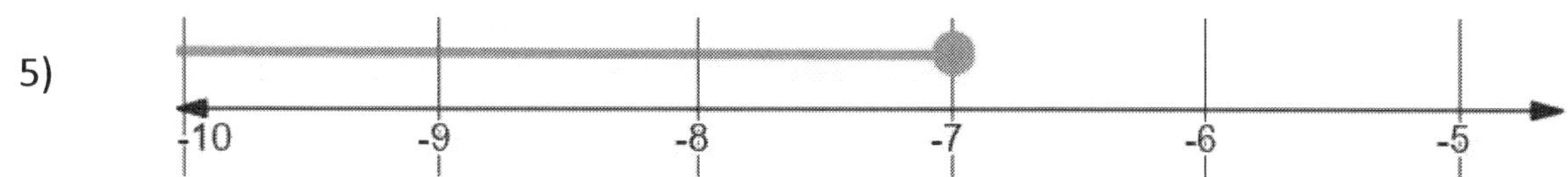

6) 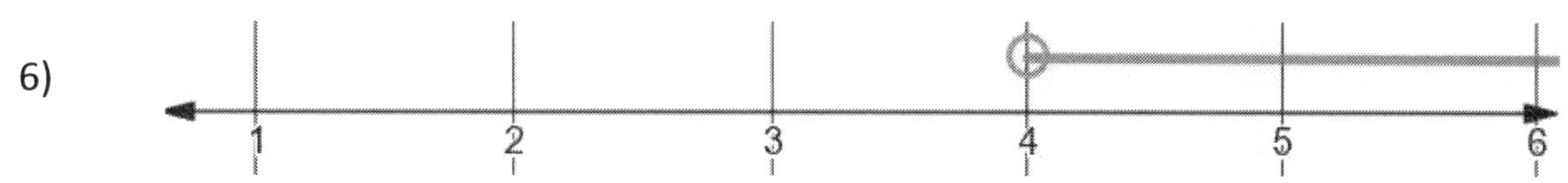

7) 

Two–Step inequalities

1) $x \leq 3$
2) $x \leq 4$
3) $x \leq 3$
4) $x \geq 2$
5) $x \geq 4$
6) $x \leq 5$
7) $x \leq 1$
8) $x \leq 1$
9) $x \leq 1$
10) $x \leq 1$
11) $x < 9.5$
12) $x < 12$
13) $x \geq 3$
14) $x < 3$
15) $x \geq 8$
16) $x < 4$

Multi–Step inequalities

1) $x < 7$
2) $x \leq 4$
3) $x > 5$
4) $x < 0$
5) $x < 9$
6) $x \leq 17$

Test Preparation Answers

1) Choice A is correct

Let x be the number of new shoes the team can purchase. Therefore, the team can purchase $120\,x$.

The team had $20,000 and spent $14000. Now the team can spend on new shoes $6000 at most.

Now, write the inequality:

$120x + 14.000 \ \leq 20.000$

2) Choice A is correct

Let x be the number of years. Therefore, $2,000 per year equals $2000x$.

starting from $24,000 annual salary means you should add that amount to $2000x$.

Income more than that is:

I > $2000x$ + 24000

3) Choice D is correct

$-2 \leq 2x - 4 < 8 \ \rightarrow$ (add 4 all sides) $-2 + 4 \leq 2x < 8 + 4 \ \rightarrow 2 \leq 2x < 12$

$\rightarrow$ (divide all sides by 2) $1 \leq x < 6$

Chapter 8: Exponents and Radicals

Topics that you'll learn in this chapter:

- ✓ Multiplication Property of Exponents
- ✓ Division Property of Exponents
- ✓ Powers of Products and Quotients
- ✓ Zero and Negative Exponents
- ✓ Negative Exponents and Negative Bases
- ✓ Writing Scientific Notation
- ✓ Square Roots

Multiplication Property of Exponents

Helpful Hints

Exponents rules

$x^a \,.\, x^b = x^{a+b}$ $\quad x^a/x^b = x^{a-b}$

$1/x^b = x^{-b}$ $\quad (x^a)^b = x^{a.b}$

$(xy)^a = x^a \,.\, y^a$

Example:

$(x^2y)^3 = x^6y^3$

Simplify.

1) $4^2 \,.\, 4^2$
2) $2 \,.\, 2^2 \,.\, 2^2$
3) $3^2 \,.\, 3^2$
4) $3x^3 \,.\, x$
5) $12x^4 \,.\, 3x$
6) $6x \,.\, 2x^2$
7) $5x^4 \,.\, 5x^4$
8) $6x^2 \,.\, 6x^3y^4$
9) $7x^2y^5 \,.\, 9xy^3$
10) $7xy^4 \,.\, 4x^3y^3$
11) $(2x^2)^2$
12) $3x^5y^3 \,.\, 8x^2y^3$
13) $7x^3 \,.\, 10y^3x^5 \,.\, 8yx^3$
14) $(x^4)^3$
15) $(2x^2)^4$
16) $(x^2)^3$
17) $(6x)^2$
18) $3x^4y^5 \,.\, 7x^2y^3$

Division Property of Exponents

Helpful Hints

$\frac{x^a}{x^b} = x^{a-b}$, $x \neq 0$

Example:

$\frac{x^{12}}{x^5} = x^7$

Simplify.

1) $\frac{5^5}{5}$

2) $\frac{3}{3^5}$

3) $\frac{2^2}{2^3}$

4) $\frac{2^4}{2^2}$

5) $\frac{x}{x^3}$

6) $\frac{3x^3}{9x^4}$

7) $\frac{2x^{-5}}{9x^{-2}}$

8) $\frac{21x^8}{7x^3}$

9) $\frac{7x^6}{4x^7}$

10) $\frac{6x^2}{4x^3}$

11) $\frac{5x}{10x^3}$

12) $\frac{3x^3}{2x^5}$

13) $\frac{12x^3}{14x^6}$

14) $\frac{12x^3}{9y^8}$

15) $\frac{25xy^4}{5x^6y^2}$

16) $\frac{2x^4}{7x}$

17) $\frac{16x^2y^8}{4x^3}$

18) $\frac{12x^4}{15x^7y^9}$

19) $\frac{12yx^4}{10yx^8}$

20) $\frac{16x^4y}{9x^8y^2}$

21) $\frac{5x^8}{20x^8}$

Powers of Products and Quotients

Helpful Hints	For any nonzero numbers a and b and any integer x, $(ab)^x = a^x \cdot b^x$.	**Example:** $(2x^2 \cdot y^3)^2 =$ $4x^2 \cdot y^6$

Simplify.

1) $(2x^3)^4$

2) $(4xy^4)^2$

3) $(5x^4)^2$

4) $(11x^5)^2$

5) $(4x^2y^4)^4$

6) $(2x^4y^4)^3$

7) $(3x^2y^2)^2$

8) $(3x^4y^3)^4$

9) $(2x^6y^8)^2$

10) $(12x\ 3x)^3$

11) $(2x^9\ x^6)^3$

12) $(5x^{10}y^3)^3$

13) $(4x^3\ x^2)^2$

14) $(3x^3\ 5x)^2$

15) $(10x^{11}y^3)^2$

16) $(9x^7\ y^5)^2$

17) $(4x^4y^6)^5$

18) $(4x^4)^2$

19) $(3x\ 4y^3)^2$

20) $(9x^2y)^3$

21) $(12x^2y^5)^2$

Zero and Negative Exponents

Helpful Hints

A negative exponent simply means that the base is on the wrong side of the fraction line, so you need to flip the base to the other side. For instance, "x^{-2}" (pronounced as "ecks to the minus two") just means "x^2" but underneath, as in $\frac{1}{x^2}$

Example:

$5^{-2} = \frac{1}{25}$

Evaluate the following expressions.

1) 8^{-2}
2) 2^{-4}
3) 10^{-2}
4) 5^{-3}
5) 22^{-1}
6) 9^{-1}
7) 3^{-2}
8) 4^{-2}
9) 5^{-2}
10) 35^{-1}
11) 6^{-3}
12) 0^{15}
13) 10^{-9}
14) 3^{-4}
15) 5^{-2}
16) 2^{-3}
17) 3^{-3}
18) 8^{-1}
19) 7^{-3}
20) 6^{-2}
21) $(\frac{2}{3})^{-2}$
22) $(\frac{1}{5})^{-3}$
23) $(\frac{1}{2})^{-8}$
24) $(\frac{2}{5})^{-3}$
25) 10^{-3}
26) 1^{-10}

Negative Exponents and Negative Bases

Helpful Hints		Example:
	– Make the power positive. A negative exponent is the reciprocal of that number with a positive exponent. – The parenthesis is important! -5^{-2} is not the same as $(-5)^{-2}$ $-5^{-2} = -\frac{1}{5^2}$ and $(-5)^{-2} = +\frac{1}{5^2}$	$2x^{-3} = \frac{2}{x^3}$

Simplify.

1) -6^{-1}

2) $-4x^{-3}$

3) $-\frac{5x}{x^{-3}}$

4) $-\frac{a^{-3}}{b^{-2}}$

5) $-\frac{5}{x^{-3}}$

6) $\frac{7b}{-9c^{-4}}$

7) $-\frac{5n^{-2}}{10p^{-3}}$

8) $\frac{4ab^{-2}}{-3c^{-2}}$

9) $-12x^2y^{-3}$

10) $(-\frac{1}{3})^{-2}$

11) $(-\frac{3}{4})^{-2}$

12) $(\frac{3a}{2c})^{-2}$

13) $(-\frac{5x}{3yz})^{-3}$

14) $-\frac{2x}{a^{-4}}$

Writing Scientific Notation

Helpful Hints

– It is used to write very big or very small numbers in decimal form.

– In scientific notation all numbers are written in the form of:

$m \times 10^n$

Decimal notation	Scientific notation
5	5×10^0
–25,000	-2.5×10^4
0.5	5×10^{-1}
2,122.456	$2{,}122456 \times 10^3$

Write each number in scientific notation.

1) 91×10^3

2) 60

3) 2000000

4) 0.0000006

5) 354000

6) 0.000325

7) 2.5

8) 0.00023

9) 56000000

10) 2000000

11) 78000000

12) 0.0000022

13) 0.00012

14) 0.004

15) 78

16) 1600

17) 1450

18) 130000

19) 60

20) 0.113

21) 0.02

Square Roots

Helpful Hints		
	– A square root of x is a number r whose square is: $r^2 = x$	**Example:**
	r is a square root of x.	$\sqrt{4} = 2$

Find the value each square root.

1) $\sqrt{1}$

2) $\sqrt{4}$

3) $\sqrt{9}$

4) $\sqrt{25}$

5) $\sqrt{16}$

6) $\sqrt{49}$

7) $\sqrt{36}$

8) $\sqrt{0}$

9) $\sqrt{64}$

10) $\sqrt{81}$

11) $\sqrt{121}$

12) $\sqrt{225}$

13) $\sqrt{144}$

14) $\sqrt{100}$

15) $\sqrt{256}$

16) $\sqrt{289}$

17) $\sqrt{324}$

18) $\sqrt{400}$

19) $\sqrt{900}$

20) $\sqrt{529}$

21) $\sqrt{90}$

Test Preparation

1) What is the value of 5^4 ?

   Write your answer in the box below.

2) How is this number written in scientific notation?

   0.00002389

   A. 2.389×10^{-5}
   B. 23.89×10^{6}
   C. 0.2389×10^{-4}
   D. 2389×10^{-8}
   E. 2389×10^{8}

3) What is the value of 3^6?

   Write your answer in the box below.

4) How is this number written in scientific notation?

0.0050468

A. 5.0468×10^{-3}

B. 5.0468×10^{3}

C. 0.50468×10^{-2}

D. 50468×10^{-7}

E. 50468×10^{7}

Answers of Worksheets – Chapter 8

Multiplication Property of Exponents

1) 4^4
2) 2^5
3) 3^4
4) $3x^4$
5) $36x^5$
6) $12x^3$
7) $25x^8$
8) $36x^5y^4$
9) $63x^3y^8$
10) $28x^4y^7$
11) $4x^4$
12) $24x^7y^6$
13) $560x^{11}y^4$
14) x^{12}
15) $16x^8$
16) x^6
17) $36x^2$
18) $21x^6y^8$

Division Property of Exponents

1) 5^4
2) $\frac{1}{3^4}$
3) $\frac{1}{2}$
4) 2^2
5) $\frac{1}{x^2}$
6) $\frac{1}{3x}$
7) $\frac{2}{9x^3}$
8) $3x^5$
9) $\frac{7}{4x}$
10) $\frac{3}{2x}$
11) $\frac{1}{2x^2}$
12) $\frac{3}{2x^2}$
13) $\frac{6}{7x^3}$
14) $\frac{4x^3}{3y^8}$
15) $\frac{5y^2}{x^5}$
16) $\frac{2x^3}{7}$
17) $\frac{4y^8}{x}$
18) $\frac{4}{5x^3y^9}$
19) $\frac{6}{5x^4}$
20) $\frac{16}{9x^4y}$
21) $\frac{1}{4}$

Powers of Products and Quotients

1) $16x^{12}$
2) $16x^2y^8$
3) $25x^8$
4) $121x^{10}$
5) $256x^8y^{16}$
6) $8x^{12}y^{12}$
7) $9x^4y^4$
8) $81x^{16}y^{12}$
9) $4x^{12}y^{16}$
10) $46{,}656x^6$
11) $8x^{45}$
12) $125x^{30}y^9$
13) $16x^{10}$
14) $225x^8$
15) $100x^{22}y^6$
16) $81x^{14}y^{10}$
17) $1{,}024x^{20}y^{30}$
18) $16x^8$

19) $144x^2y^6$

20) $729x^6y^3$

21) $144x^4y^{10}$

Zero and Negative Exponents

1) $\frac{1}{64}$

2) $\frac{1}{16}$

3) $\frac{1}{100}$

4) $\frac{1}{125}$

5) $\frac{1}{22}$

6) $\frac{1}{9}$

7) $\frac{1}{9}$

8) $\frac{1}{16}$

9) $\frac{1}{25}$

10) $\frac{1}{35}$

11) $\frac{1}{216}$

12) 0

13) $\frac{1}{1000000000}$

14) $\frac{1}{81}$

15) $\frac{1}{25}$

16) $\frac{1}{8}$

17) $\frac{1}{27}$

18) $\frac{1}{8}$

19) $\frac{1}{343}$

20) $\frac{1}{36}$

21) $\frac{9}{4}$

22) 125

23) 256

24) $\frac{125}{8}$

25) $\frac{1}{1000}$

26) 1

Negative Exponents and Negative Bases

1) $-\frac{1}{6}$

2) $-\frac{4}{x^3}$

3) $-5x^4$

4) $-\frac{b^2}{a^3}$

5) $-5x^3$

6) $-\frac{7bc^4}{9}$

7) $-\frac{p^3}{2n^2}$

8) $-\frac{4ac^2}{3b^2}$

9) $-\frac{12x^2}{y^3}$

10) 9

11) $\frac{16}{9}$

12) $\frac{4c^2}{9a^2}$

13) $-\frac{27y^3z^3}{125x^3}$

14) $-2xa^4$

Writing Scientific Notation

1) 9.1×10^4

2) 6×10^1

3) 2×10^6

4) 6×10^{-7}

5) 3.54×10^5

6) 3.25×10^{-4}

7) 2.5×10^0

8) 2.3×10^{-4}

9) 5.6×10^7

10) 2×10^6
11) 7.8×10^7
12) 2.2×10^{-6}
13) 1.2×10^{-4}
14) 4×10^{-3}
15) 7.8×10^1
16) 1.6×10^3
17) 1.45×10^3
18) 1.3×10^5
19) 6×10^1
20) 1.13×10^{-1}
21) 2×10^{-2}

Square Roots

1) 1
2) 2
3) 3
4) 5
5) 4
6) 7
7) 6
8) 0
9) 8
10) 9
11) 11
12) 15
13) 12
14) 10
15) 16
16) 17
17) 18
18) 20
19) 30
20) 23
21) $3\sqrt{10}$

Test Preparation Answers

1) The answer is 625.

$5^4 = 5 \times 5 \times 5 \times 5 = 625$

2) Choice A is correct.

$0.00002389 = \frac{2.389}{100000} \Rightarrow 2.389 \times 10^{-5}$

3) The answer is 729.

$3^6 = 3 \times 3 \times 3 \times 3 \times 3 \times 3 = 729$

4) Choice A is correct

$0.0050468 = \frac{5.0468}{1000} \Rightarrow 5.0468 \times 10^{-3}$

Chapter 9: Measurements

Topics that you'll learn in this chapter:

- ✓ Inches & Centimeters
- ✓ Metric units
- ✓ Distance Measurement
- ✓ Weight Measurement

Inches and Centimeters

Helpful Hints	1 inch = 2.5 cm 1 foot = 12 inches 1 yard = 3 feet 1 yard = 36 inches 1 inch = 0.0254 m	**Example:** 18 inches = 0.4572 m

Convert to the units.

1inch = 2.5 cm

1) 25 cm = _____ inches

2) 11 inches = _____ cm

3) 1 m = _____ inches

4) 80 inches = _____ m

5) 200 cm = _____ m

6) 5 m = _____ cm

7) 4 feet = ____ inches

8) 10 yards = ____ inches

9) 16 feet = ____ inches

10) 48 inches = ____ Feet

11) 4 inches = _____ cm

12) 12.5 cm = _____ inches

13) 6 feet: ____ inches

14) 10 feet: ____ inches

15) 12 yards: ____ feet

16) 7 yards: ____ feet

Metric Units

Helpful Hints		Example:
	1 m = 100 cm	
	1 cm = 10 mm	
	1 m = 1000 mm	12 cm = 0.12 m
	1 km = 1000 m	

Convert to the units.

1) 4 mm = ______ cm

2) 0.6 m = ______ mm

3) 2 m = ______ cm

4) 0.03 km = ______ m

5) 3000 mm = ______ km

6) 5 cm = ______ m

7) 0.03 m = ______ cm

8) 1000 mm = ______ km

9) 600 mm = ______ m

10) 0.77 km = ______ mm

11) 0.08 km = ______ m

12) 0.30 m = ______ cm

13) 400 m = ______ km

14) 5000 cm = ______ km

15) 40 mm = ______ cm

16) 800 m = ______ km

Distance Measurement

Helpful Hints

1 mile = 5280 ft

1 mile = 1760 yd

1 mile = 1609.34 m

Example:

10 miles = 52800 ft

Convert to the new units.

1) 2 mi = _______ ft

2) 21 mi = _______ ft

3) 6 mi = _______ ft

4) 3 mi = _______ yd

5) 72 mi = _______ ft

6) 41 mi = _______ yd

7) 62 mi = _______ yd

8) 39 mi = _______ yd

9) 7 mi = _______ yd

10) 94 mi = _______ yd

11) 87 mi = _______ yd

12) 23 mi = _______ yd

13) 2 mi = _______ m

14) 5 mi = _______ m

15) 6 mi = _______ m

16) 3 mi = _______ m

Weight Measurement

Helpful Hints	1 kg = 1000g	**Example:** 2000 g = 2 kg

Convert to grams.

1) 0.5 kg = __________ g
2) 3.2 kg = __________ g
3) 8.2 kg = __________ g
4) 9.2 kg = __________ g
5) 35 kg = __________ g
6) 87 kg = __________ g
7) 45 kg = __________ g
8) 15 kg = __________ g
9) 0.32 kg = __________ g
10) 81 kg = __________ g

Convert to kilograms.

11) 200,000 g = _______ kg
12) 30,000 g = _______ kg
13) 800,000 g = _______ kg
14) 20,000 g = _______ kg
15) 40,000 g = _______ kg
16) 500,000 g = _______ kg

Test Preparation

1) 12 yards 4 feet and 2 inches equals to how many inches?

A. 96
B. 432
C. 482
D. 578
E. 620

2) A rope weighs 800 grams per meter of length. What is the weight in kilograms of 12.2 meters of this rope? (1 kilograms = 1000 grams)

A. 0.0976
B. 0.976
C. 9.76
D. 9,760
E. 9,990

3) A house floor has a perimeter of 4,221 feet. What is the perimeter of the floor in yards?

Write your answer in the box below.

Answers of Worksheets – Chapter 9

Inches & Centimeters

1) 25 cm = 9.84 inches
2) 11 inch = 27.94 cm
3) 1 m = 39.37 inches
4) 80 inch = 2.03 m
5) 200 cm = 2 m
6) 5 m = 500 cm
7) 4 feet = 48 inches
8) 10 yards = 360 inches
9) 16 feet = 192 inches
10) 48 inches = 4 Feet
11) 4 inch = 10.16 cm
12) 12.5 cm = 4.92 inches
13) 6 feet: 72 inches
14) 10 feet: 120 inches
15) 12 yards: 36 feet
16) 7 yards: 21 feet

Metric Units

1) 4 mm = 0.4 cm
2) 0.6 m = 600 mm
3) 2 m = 200 cm
4) 0.03 km = 30 m
5) 3000 mm = 0.003 km
6) 5 cm = 0.05 m
7) 0.03 m = 3 cm
8) 1000 mm = 0.001 km
9) 600 mm = 0.6 m
10) 0.77 km = 770,000 mm
11) 0.08 km = 80 m
12) 0.30 m = 30 cm
13) 400 m = 0.4 km
14) 5000 cm = 0.05 km
15) 40 mm = 4 cm
16) 800 m = 0.8 km

Distance Measurement

1) 21 mi = 110880 ft
2) 6 mi = 31680 ft
3) 3 mi = 5280 yd
4) 72 mi = 380160 ft
5) 41 mi = 72160 yd
6) 62 mi = 109120 yd
7) 39 mi = 68640 yd
8) 7 mi = 12320 yd
9) 94 mi = 165440 yd
10) 87 mi = 153120 yd

11) 23 mi = 40480 yd
12) 2 mi = 3218.69 m
13) 5 mi = 8046.72 m
14) 6 mi = 9656.06 m
15) 3 mi = 4828.03 m

Weight Measurement

1) 0.5 kg = 500 g
2) 3.2 kg = 3200 g
3) 8.2 kg = 8200 g
4) 9.2 kg = 9200 g
5) 35 kg = 35000 g
6) 87 kg = 87000 g
7) 45 kg = 45000 g
8) 15 kg = 15000 g
9) 0.32 kg = 320 g
10) 81 kg = 81000 g
11) 200,000 g = 200 kg
12) 30,000 g = 30 kg
13) 800,000 g = 800 kg
14) 20,000 g = 20 kg
15) 40,000 g = 40 kg
16) 500,000 g = 500 kg

Test Preparation Answers

1) Choice C is correct

12Yards = (12×36) 432 inches

4feet = (4×12) 48 inches

12 yards + 4 feet + 2 inches = 432 + 48 +2 = 482

2) Choice C is correct

The weighs of rope per meter of length = 800 grams. We use ratio to find answer (Let x be the amount of weight):

$\frac{800 \text{ grams}}{1\, meter} = \frac{x \text{ grams}}{12.2\ \ meter}$ ⇒ x= 9760 grams x= 9.760 kilograms

3) The answer is 1407.

1 yard=3feet

$4{,}221 \div 3 = 1{,}407$

Chapter 10: Plane Figures

Topics that you'll learn in this chapter:

- ✓ The Pythagorean Theorem
- ✓ Area of Triangles
- ✓ Perimeter of Polygons
- ✓ Area and Circumference of Circles
- ✓ Area of Squares, Rectangles, and Parallelograms
- ✓ Area of Trapezoids

The Pythagorean Theorem

Helpful Hints

– In any right triangle:

$a^2 + b^2 = c^2$

Example:

Missing side = 6

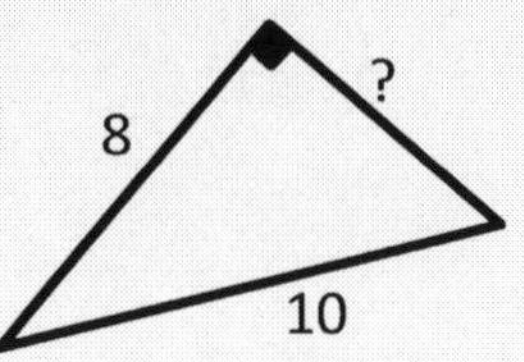

Do the following lengths form a right triangle?

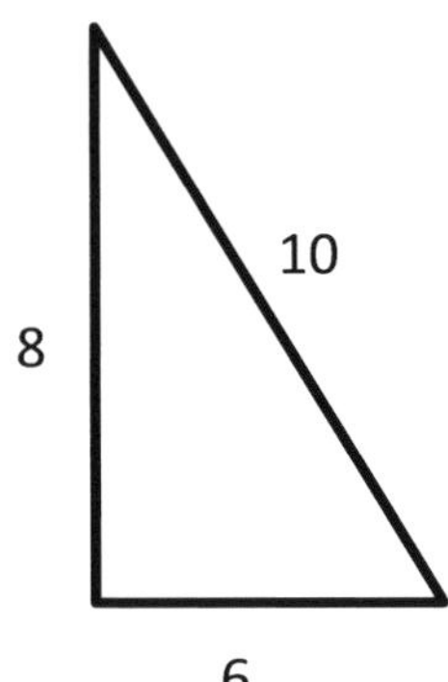

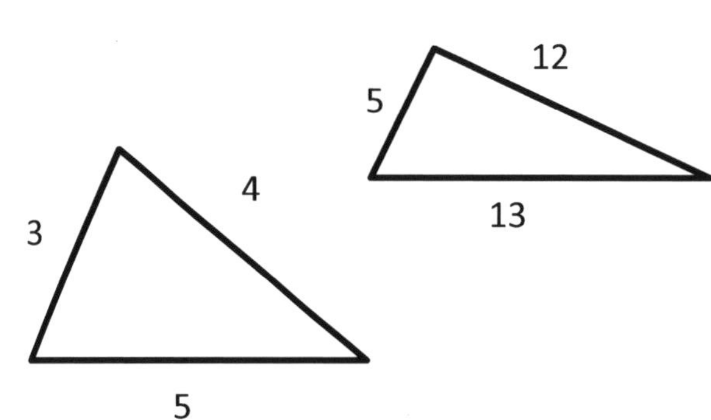

Find each missing length to the nearest tenth.

4)

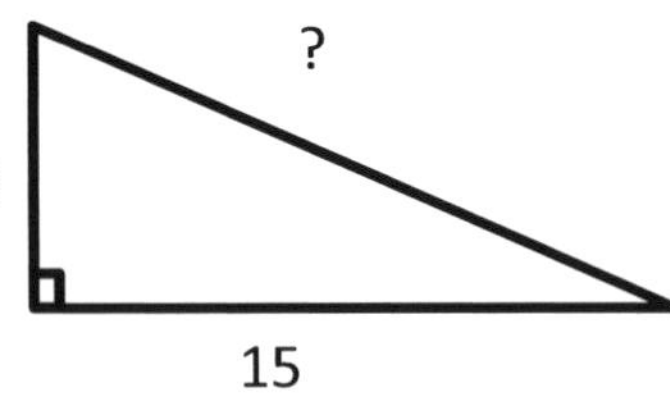

5)

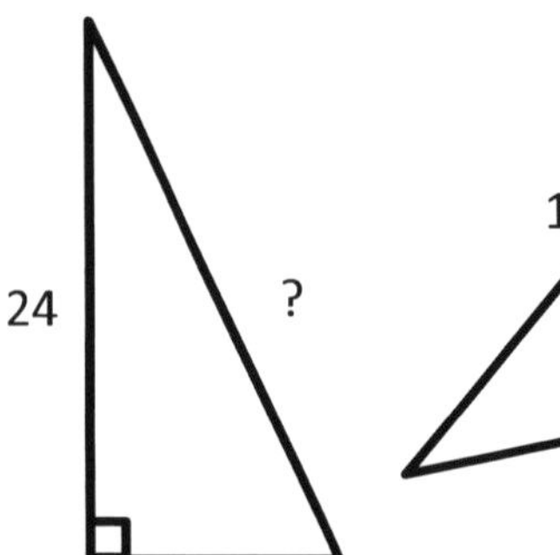

6)

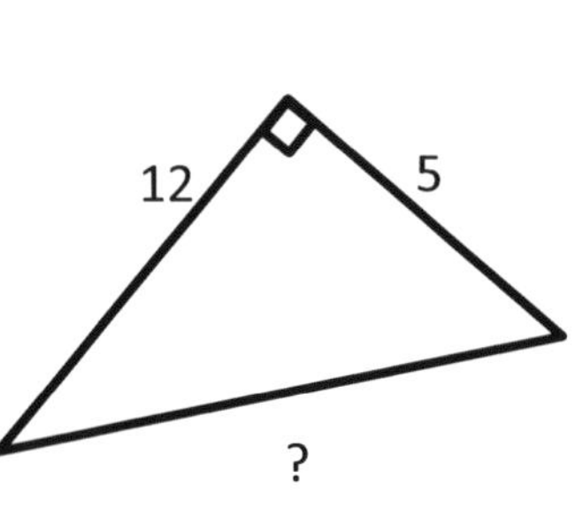

Area of Triangles

Helpful Hints

Area $= \frac{1}{2}$ $(base \times height)$

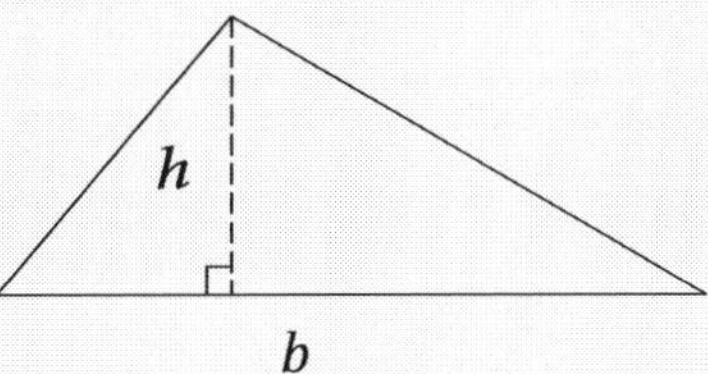

Find the area of each.

1)

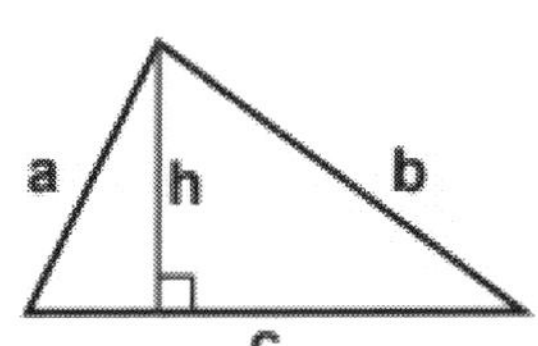

c = 9 mi

h = 3.7 mi

2)

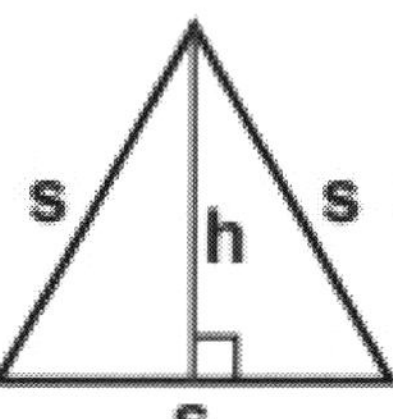

s = 14 m

h = 12.2 m

3)

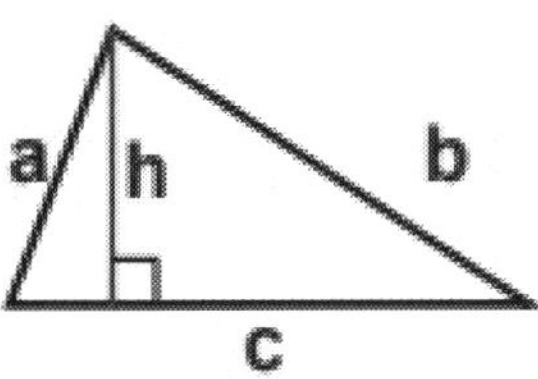

a = 5 m

b = 11 m

c = 14 m

h = 4 m

4)

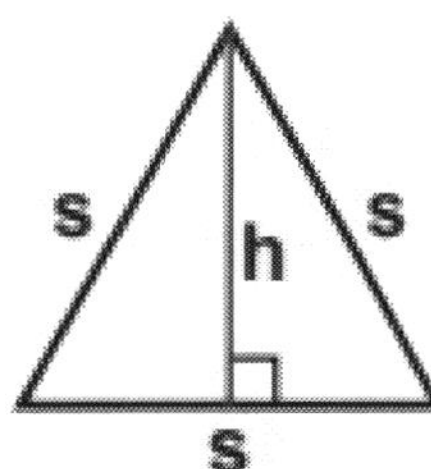

s = 10 m

h = 8.6 m

Perimeter of Polygons

Helpful Hints

Perimeter of a square = 4s

Perimeter of a rectangle

$= 2(l + w)$

Perimeter of trapezoid

= a + b + c + d

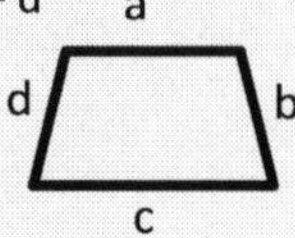

Perimeter of Pentagon = 6a

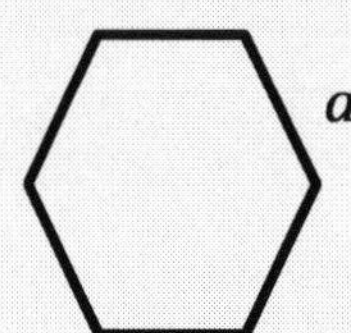

Perimeter of a parallelogram = 2(l + w)

Example:

P = 18

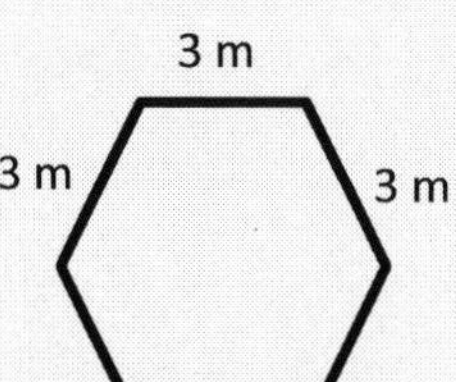

Find the perimeter of each shape.

1)

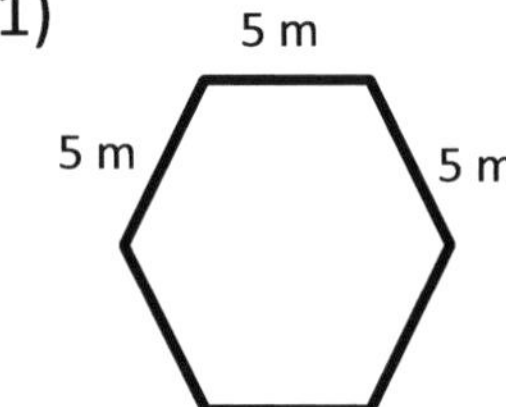

2)

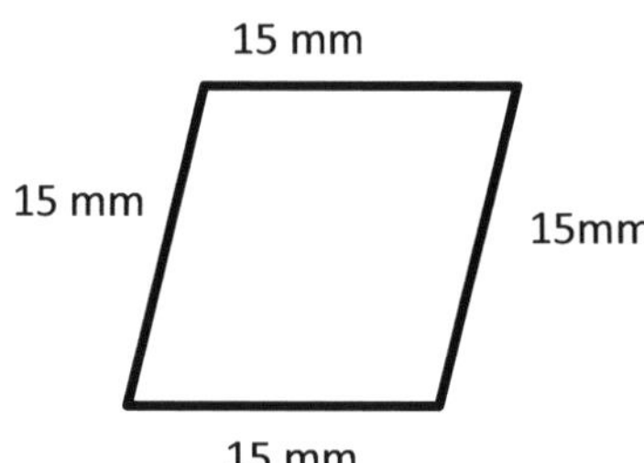

3)

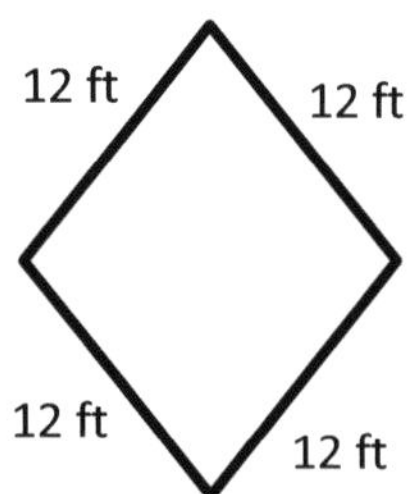

4)

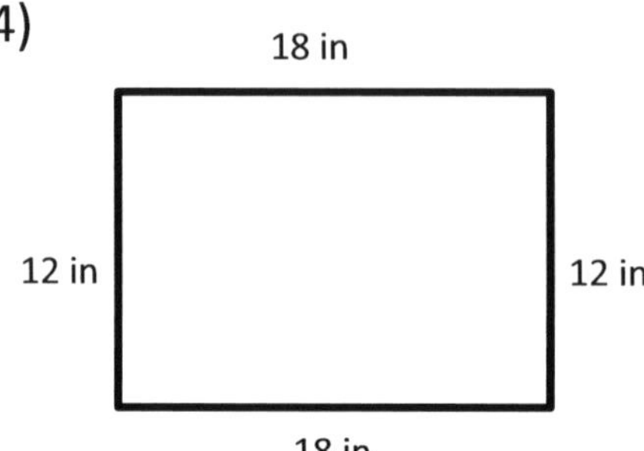

Area and Circumference of Circles

Helpful Hints

Area = πr^2

Circumference = $2\pi r$

r

Example:

If the radius of a circle is 3, then:

Area = 28.27

Circumference = 18.85

Find the area and circumference of each. $(\pi = 3.14)$

1)

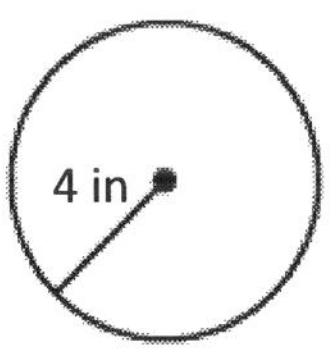

2)

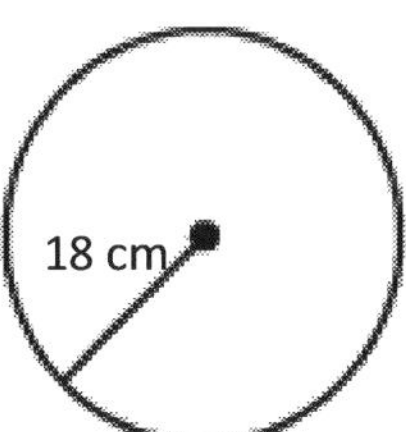

3)

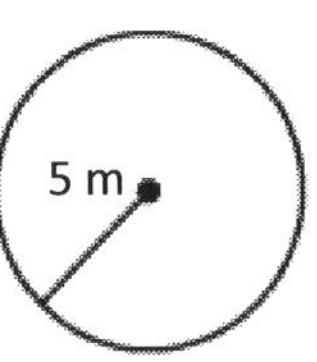

4)

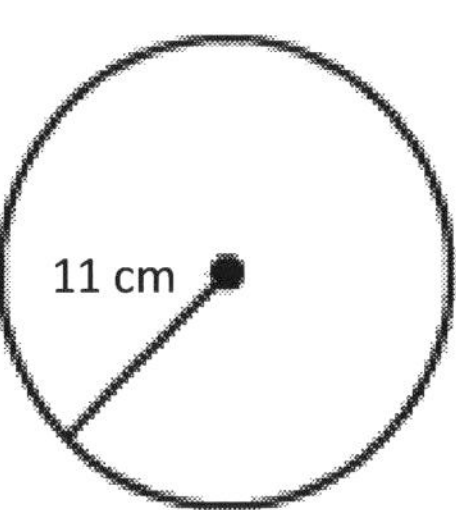

5)

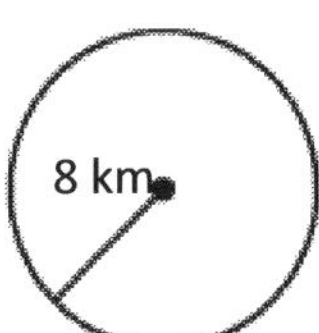

6)

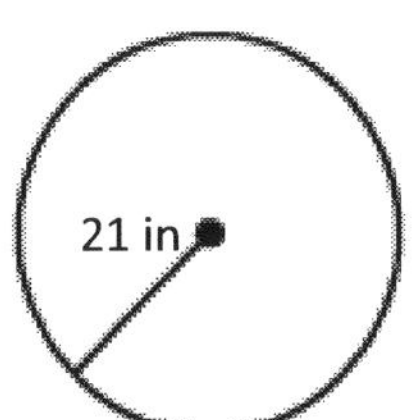

Area of Squares, Rectangles, and Parallelograms

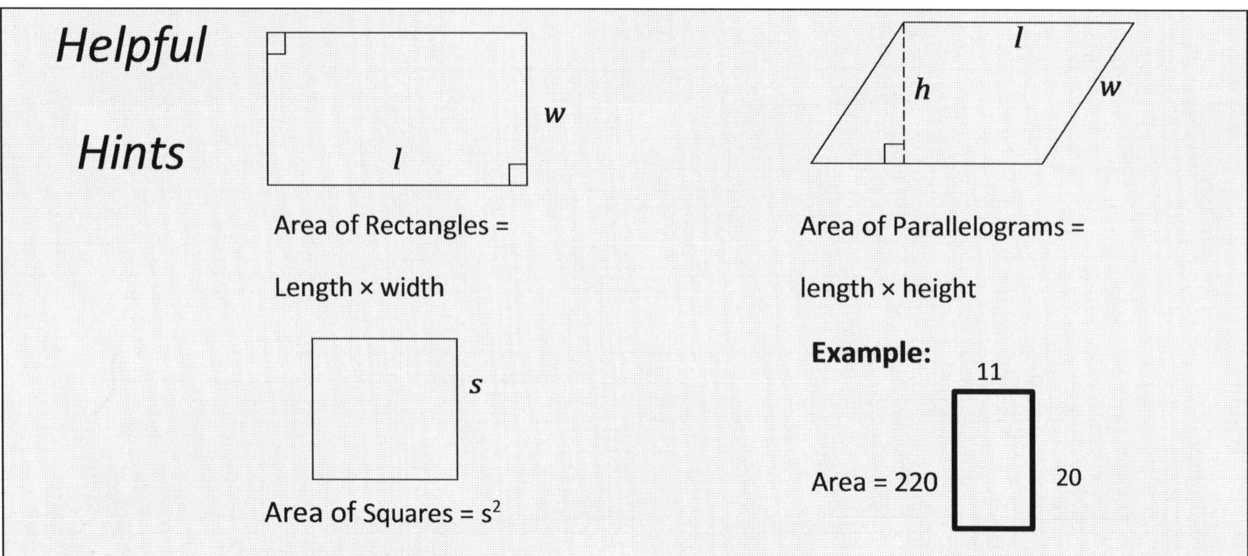

Find the area of each.

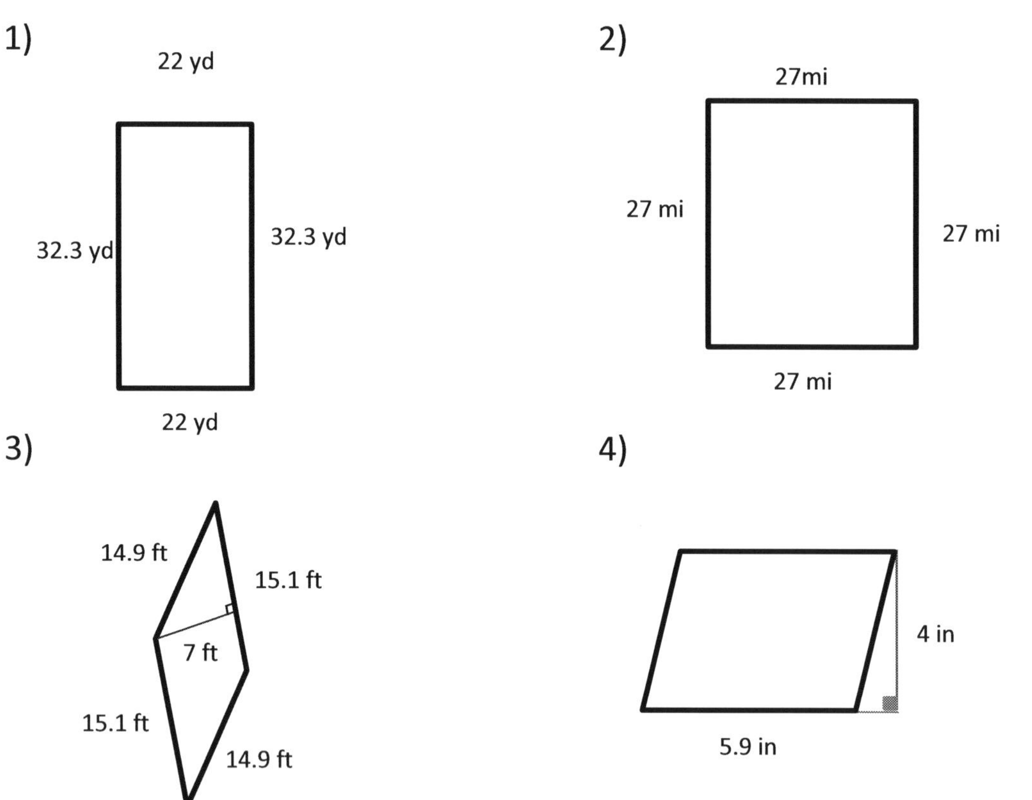

Area of Trapezoids

Helpful Hints

$A = \frac{1}{2}h(b_1 + b_2)$

Example:

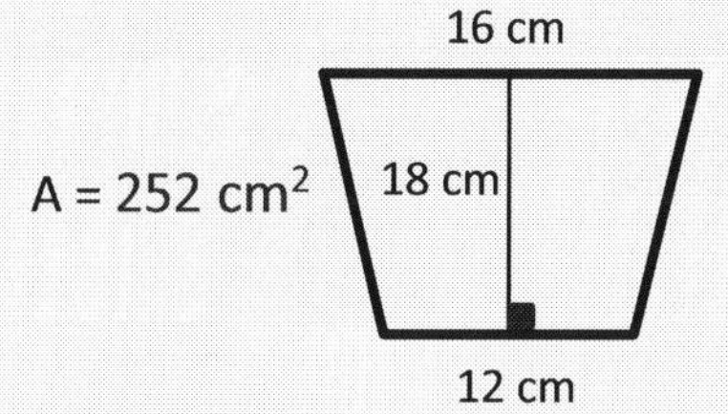

Calculate the area for each trapezoid.

1)

9 cm

6 cm

12 cm

2)

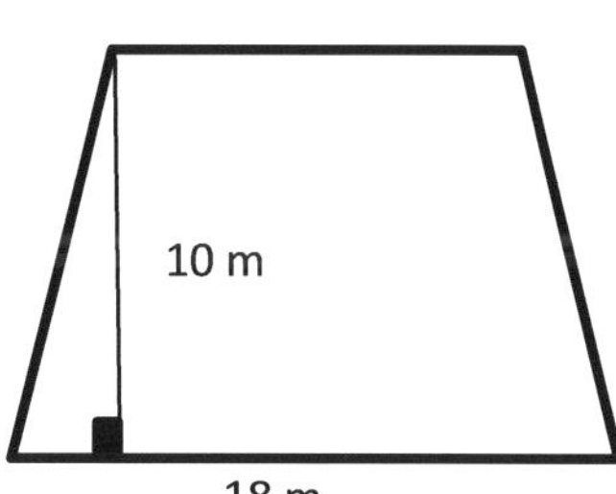

3)

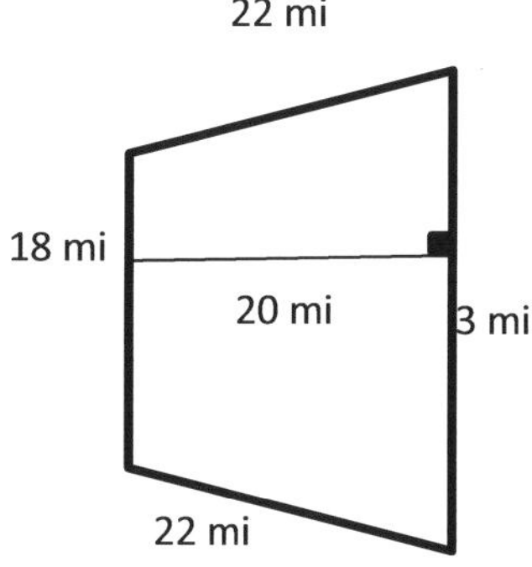

4)

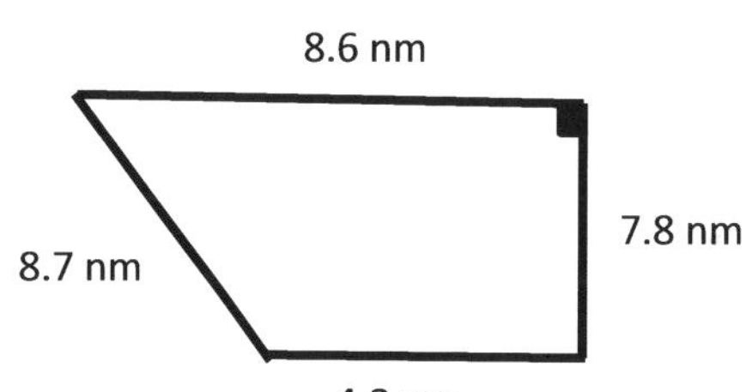

Test Preparation

1) In a triangle ABC the measure of angle ACB is 68° and the measure of angle CAB is 52°. What is the measure of angle ABC?

Write your answer in the box below.

2) The perimeter of a rectangular yard is 60 meters. What is its length if its width is 10 meters?

A. 10 meters

B. 18 meters

C. 20 meters

D. 24 meters

E. 36 meters

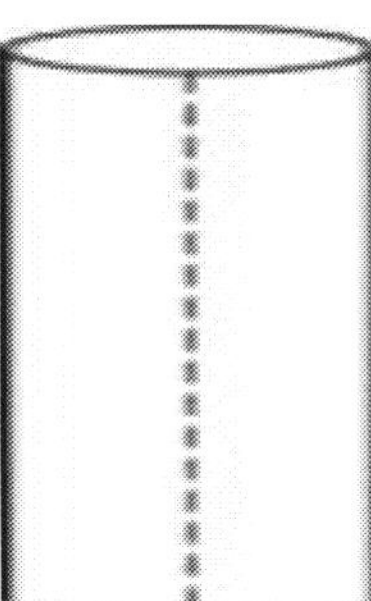

3) The radius of the following cylinder is 4 inches and its height is 10 inches. What is the volume of the cylinder?

Write your answer in the box below. (π equals 3.14) (Round your answer to the nearest whole number)

4) The perimeter of the trapezoid below is 52. What is its area?

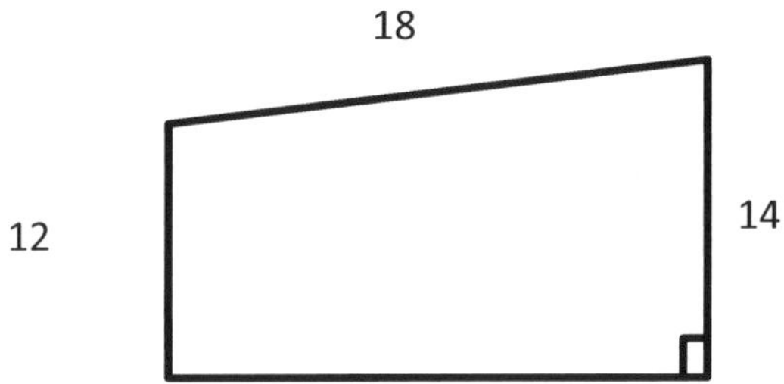

Write your answer in the box below.

5) The area of a circle is 64 π. What is the circumference of the circle?

(A) 8 π

(B) 16 π

(C) 32 π

(D) 64 π

(E) 68 π

6) The length of a rectangle is 12 inches long and its area is 96 square inches. What is the perimeter of the rectangle?

Write your answer in the box below.

7) The perimeter of the trapezoid below is 36 cm. What is its area?

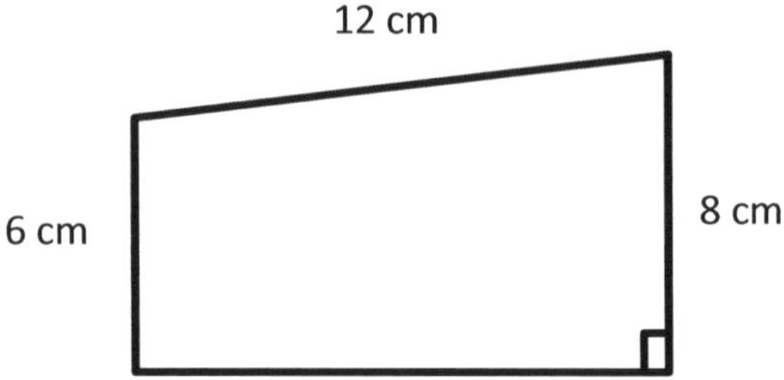

A. 26
B. 42
C. 48
D. 70
E. 80

8) What is the perimeter of a square that has an area of 169 square inches?

Write your answer in the box below.

Answers of Worksheets – Chapter 10

The Pythagorean Theorem

1) yes
2) yes
3) yes
4) 17
5) 26
6) 13

Area of Triangles

1) 16.65 mi^2
2) 85.4 m^2
3) 28 m^2
4) 43 m^2

Perimeter of Polygons

1) 30 m
2) 60 mm
3) 48 ft
4) 60 in

Area and Circumference of Circles

1) Area: 50.24 in^2, Circumference: 25.12 in
2) Area: 1,017.36 cm^2, Circumference: 113.04 cm
3) Area: 78.5m^2, Circumference: 31.4 m
4) Area: 379.94 cm^2, Circumference: 69.08 cm
5) Area: 200.96 km^2, Circumference: 50.2 km
6) Area: 1,384.74 km^2, Circumference: 131.88 km

Area of Squares, Rectangles, and Parallelograms

1) 710.6 yd^2
2) 729 mi^2
3) 105.7 ft^2
4) 23.6 in^2

Area of Trapezoids

1) 63 cm^2
2) 160 m^2
3) 410 mi^2
4) 50.31 nm^2

Test Preparation Answers

1) The answer is 60.

The whole angles in every triangle are: 180°and Let x be the number of new angle so:

180 = 68 + 52 + $x \Rightarrow x = 60°$

2) Choice C is correct

Let x be the length, and its $width = 10$

Perimeter of the rectangle is 2(width + length) = $2(10 + x) = 60 \Rightarrow 10 + x = 30 \Rightarrow x = 20$

Length of the rectangle is 10 meters.

3) The answer is 502.

The volume of the cylinder: $\pi r^2 h$

The volume of the cylinder: (3.14) × $(4)^2$× 10 = 502.4 ≅ 502

4) The answer is 104.

The perimeter of the trapezoid = the sum of the lengths of its four sides

So: 52 = 12+ 18+ 14+ $x \Rightarrow x$ =8

The area of the trapezoid = the sum of its bases, multiply the half of its height

So The area of the trapezoid = (12+14) × $\frac{8}{2}$ = 104

5) Choice B is correct

Use the formula of areas of circles.

Area = $\pi r^2 \Rightarrow 64\pi = \pi r^2 \Rightarrow 64 = r^2 \Rightarrow r = 8$

Radius of the circle is 8. Now, use the circumference formula:

Circumference = $2\pi r = 2\pi (8) = 16\pi$

6) The answer is 40.

Use the formula of areas of rectangles.

Area: length plus width $\Rightarrow 96 = 12 \times$ width $\Rightarrow$ width = 8

Use the formula of perimeter of rectangles.

Perimeter: 2(length + width) $\Rightarrow 2(12 + 8) = 40$

7) Choice D is correct

The perimeter of the trapezoid is 36 cm.

Therefore, the missing side (height) is = $36 - 8 - 12 - 6 = 10$

Area of a trapezoid: $A = \frac{1}{2} h (b_1 + b_2) = \frac{1}{2} (10)(6 + 8) = 70$

8) The answer is 52.

Use the area of a square formula.

$S = a^2 \Rightarrow 169 = a^2 \Rightarrow a = 13$

Use the perimeter of a square formula.

$P = 4a \Rightarrow p = 4 \times 13 \Rightarrow p = 52$

Chapter 11: Solid Figures

Topics that you'll learn in this chapter:

- ✓ Volume of Cubes
- ✓ Volume of Rectangle Prisms
- ✓ Surface Area of Cubes
- ✓ Surface Area of Rectangle Prisms

Volume of Cubes

Helpful	– Volume is the measure of the amount of space inside of a solid figure, like a cube, ball, cylinder or pyramid.
Hints	– Volume of a cube = $(\text{one side})^3$
	– Volume of a rectangle prism: Length × Width × Height

Find the volume of each.

1) 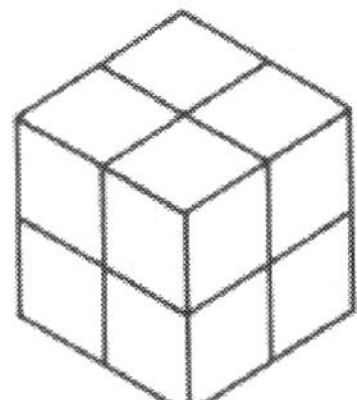

2) 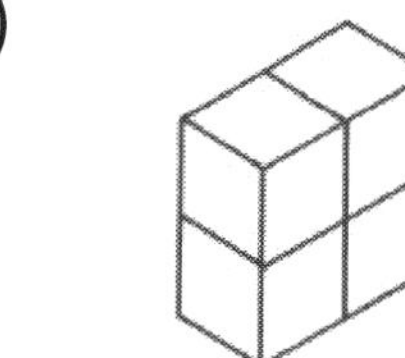

3) 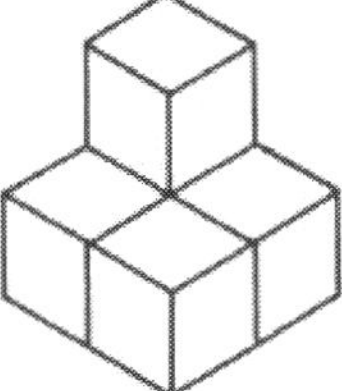

4) 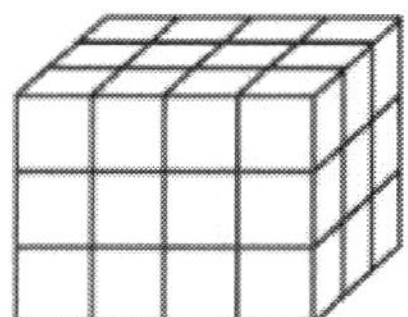

5) 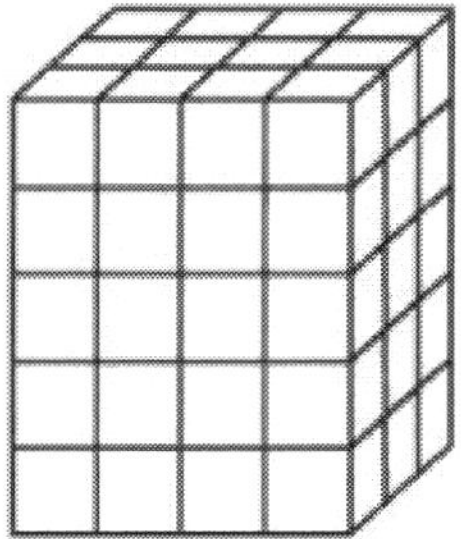

6) 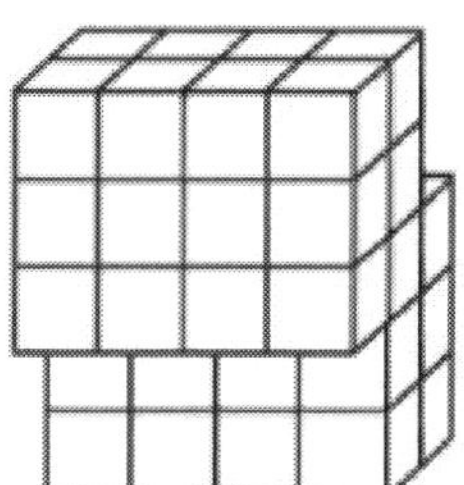

Volume of Rectangle Prisms

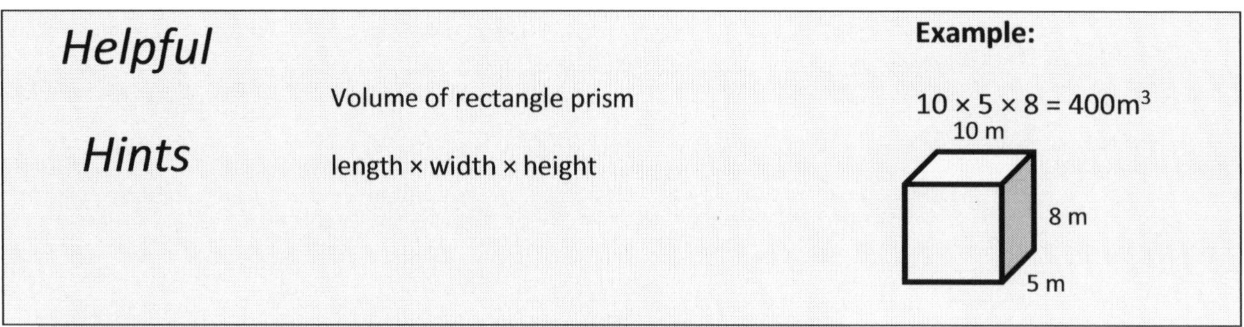

Find the volume of each of the rectangular prisms.

1)

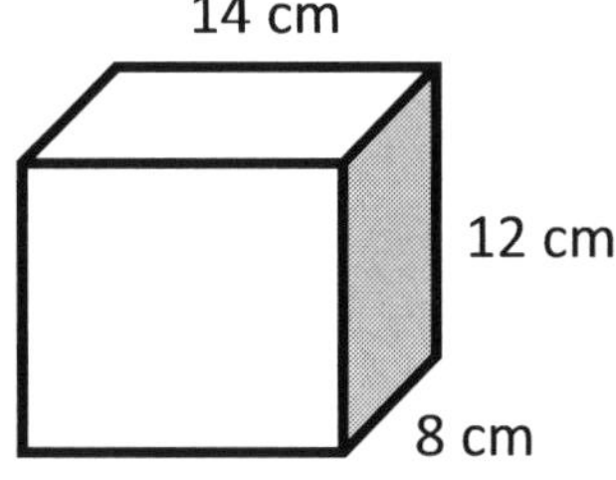

2)

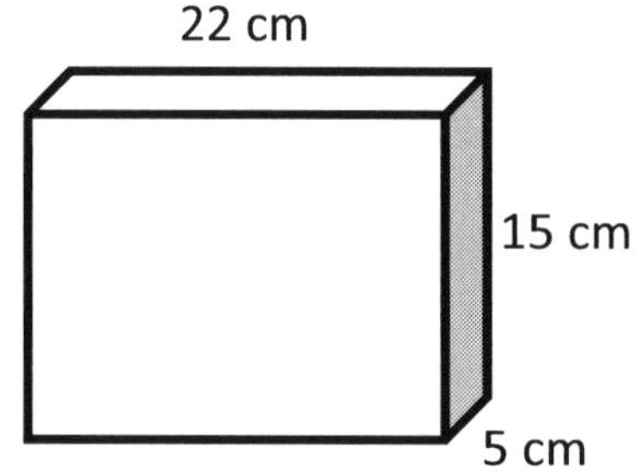

3)

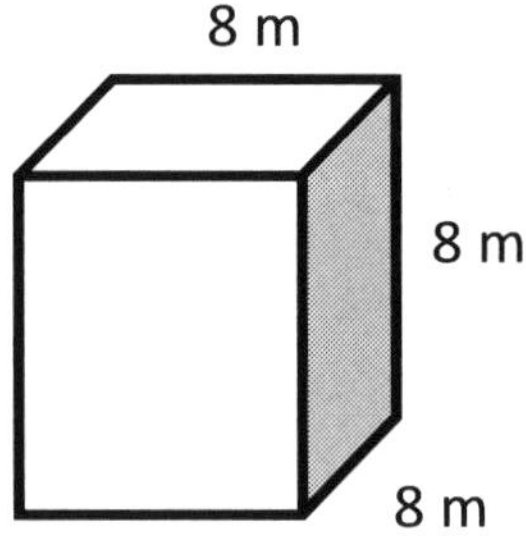

4)

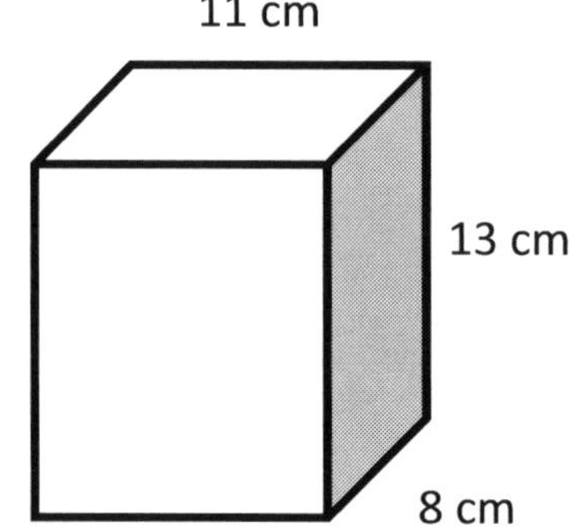

Surface Area of Cubes

Helpful Hints

Surface Area of a cube =

6 × (one side of the cube)^2

Example:

$6 \times 4^2 = 96m^2$

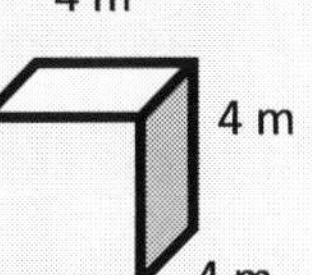

Find the surface of each cube.

1)

6 mm

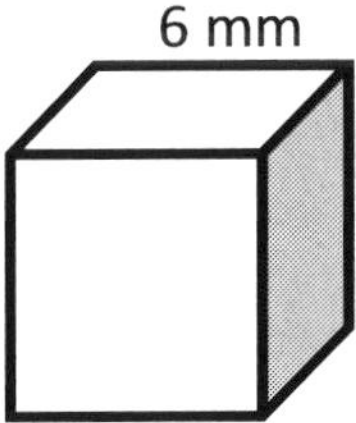

2)

9 mm

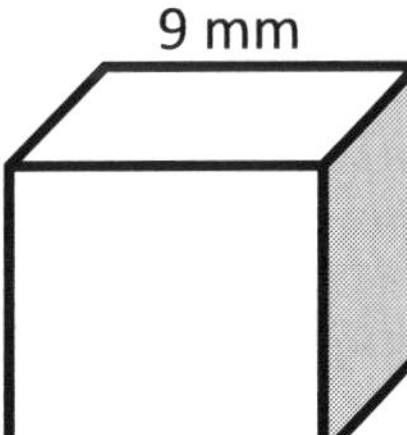

3)

10 cm

4)

8 m

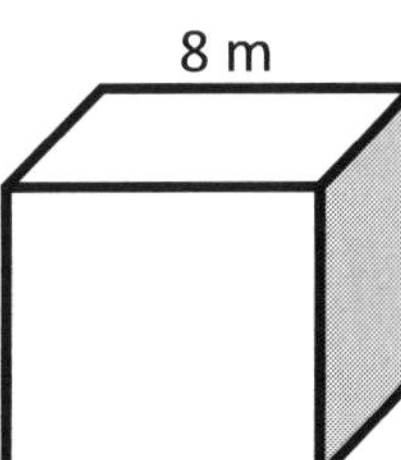

5)

7.5 in

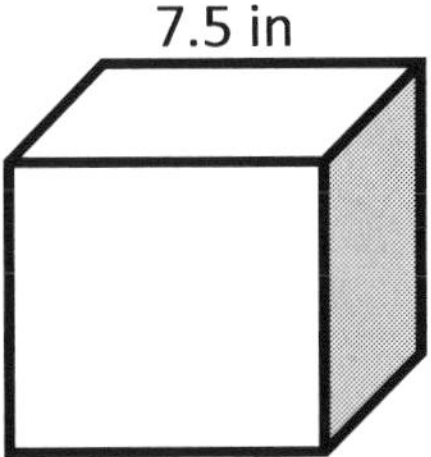

6)

11.3 ft

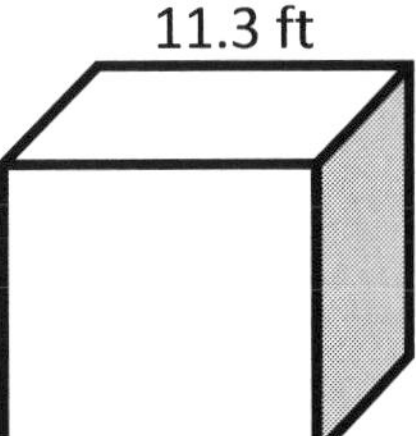

Surface Area of a Rectangle Prism

Helpful Hints

Surface Area of a Rectangle Prism Formula:

SA =2 [(width × length) + (height × length) + width × height)]

Find the surface of each prism.

1)

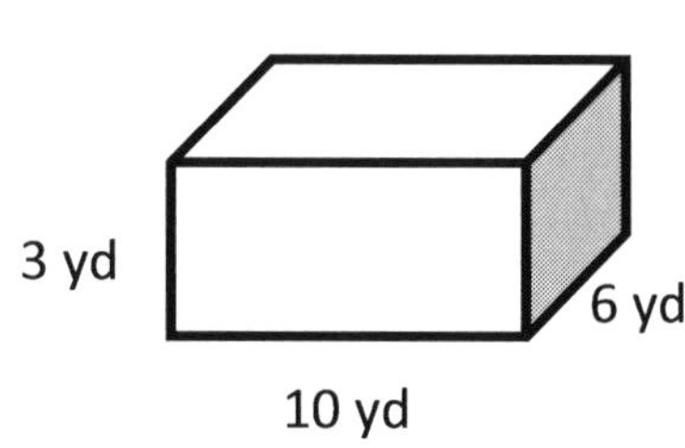

2)

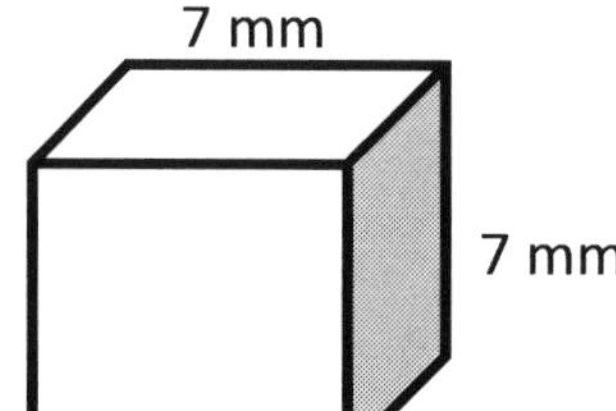

3)

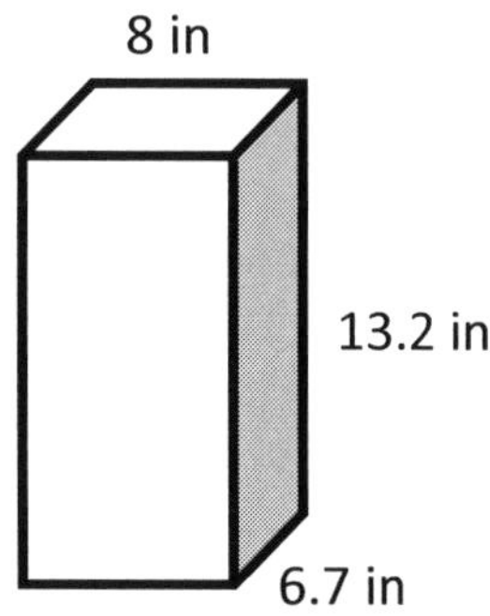

4)

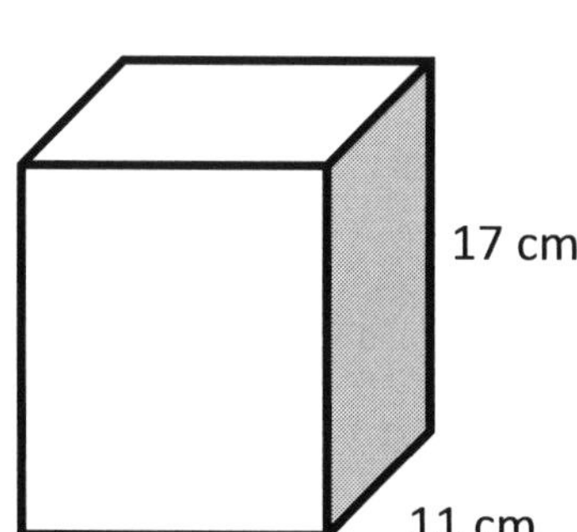

Volume of a Cylinder

Helpful Hints

Volume of Cylinder Formula = $\pi(\text{radius})^2 \times \text{height}$

$\pi = 3.14$

Find the volume of each cylinder. ($\pi = 3.14$)

1)

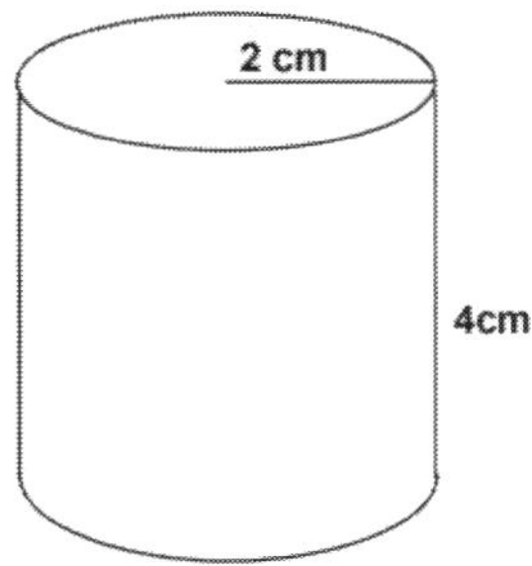

2)

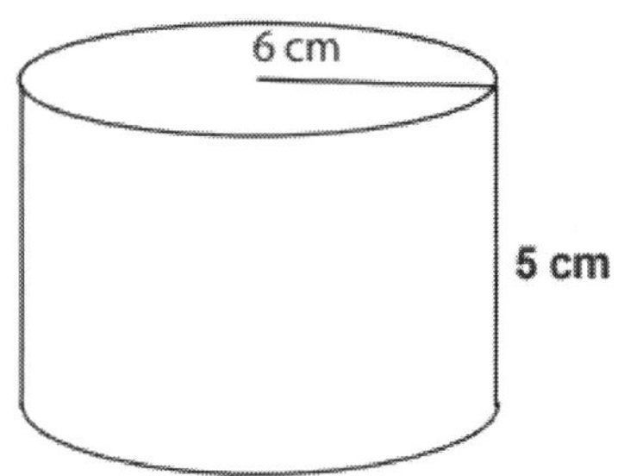

3)

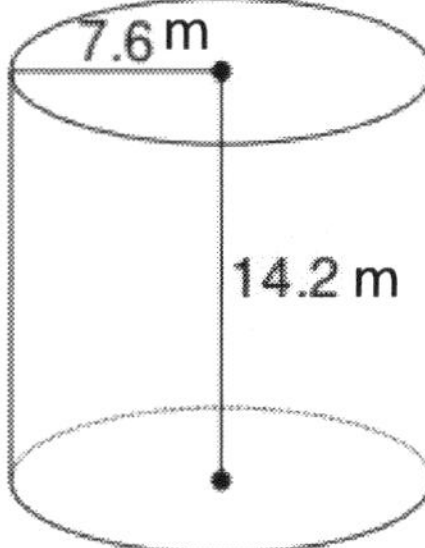

4)

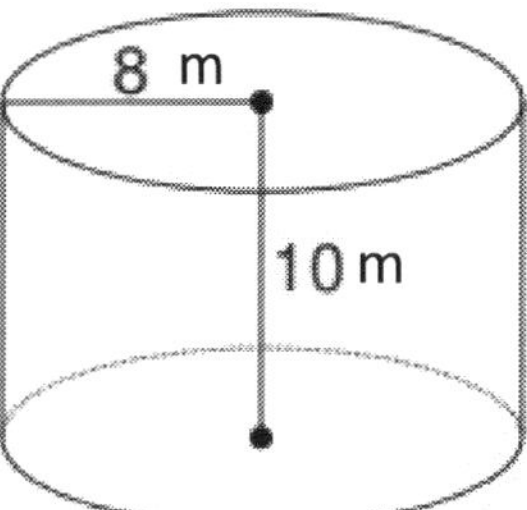

Surface Area of a Cylinder

Helpful Hints

Surface area of a cylinder

SA = $2\pi r^2 + 2\pi rh$

Example:

Surface area

= 1727

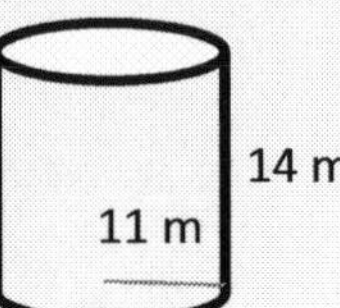

***Find the surface of each cylinder.*** $(\pi = 3.14)$

1)

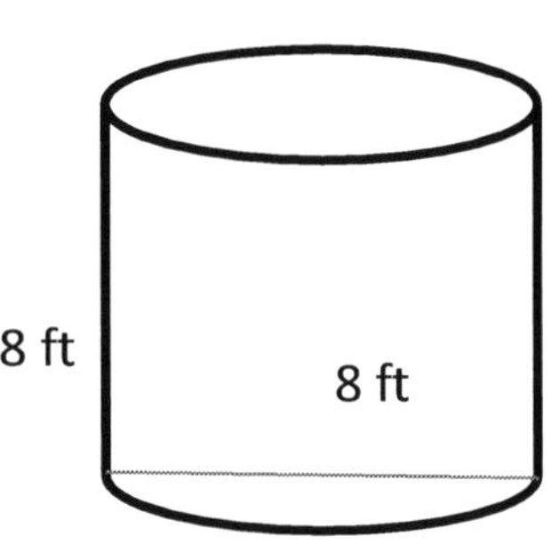

2)

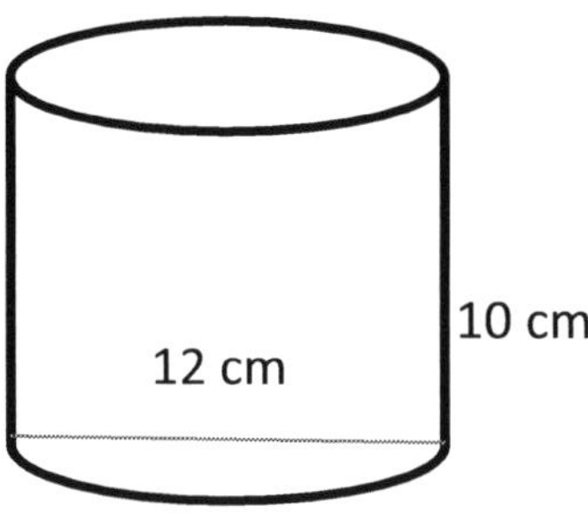

3)

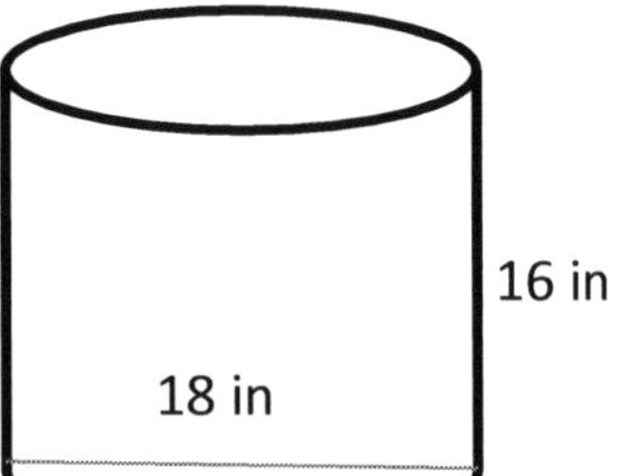

4)

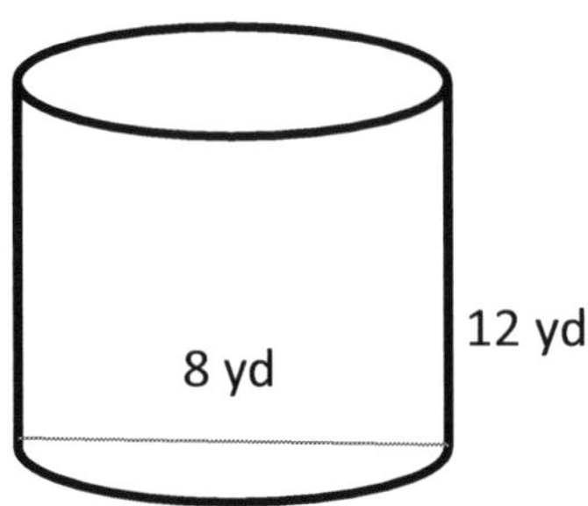

Test Preparation

1) A swimming pool holds 2,000 cubic feet of water. The swimming pool is 25 feet long and 10 feet wide. How deep is the swimming pool?

   Write your answer in the box below. (Don't write the measurement)

2) What is the volume of a cube whose side is 4 cm?

   (A) 16 cm^3

   (B) 32 cm^3

   (C) 36 cm^3

   (D) 64 cm^3

   (E) 72 cm^3

3) What is the volume of the cylinder below?

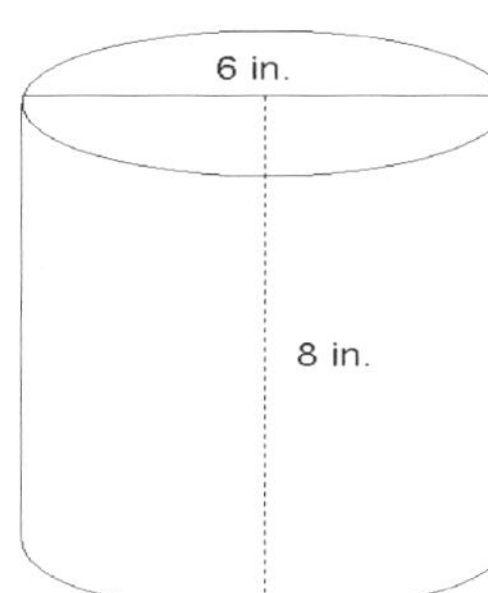

   A. 48 π in^2

   B. 57 π in^2

   C. 66 π in^2

   D. 72 π in^2

   E. 76 π in^2

4) What is the volume of a box with the following dimensions?

   Hight = 4 cm   Width = 5 cm   Length = 6 cm

   A. 15 cm^3
   B. 60 cm^3
   C. 90 cm^3
   D. 120 cm^3
   E. 160 cm^3

Answers of Worksheets – Chapter 11

Volumes of Cubes

1) 8
2) 4
3) 5
4) 36
5) 60
6) 44

Volume of Rectangle Prisms

1) 1344 cm^3
2) 1650 cm^3
3) 512 m^3
4) 1144 cm^3

Surface Area of a Cube

1) 216 mm^2
2) 486 mm^2
3) 600 cm^2
4) 384 m^2
5) 337.5 in^2
6) 766.14 ft^2

Surface Area of a Prism

1) 216 yd^2
2) 294 mm^2
3) 495.28 in^2
4) 1326 cm^2

Volume of a Cylinder

1) 50.24 cm^3
2) 565.2 cm^3
3) $2,575.403 \text{ m}^3$
4) 2009.6 m^3

Surface Area of a Cylinder

1) 301.44 ft^2
2) 602.88 cm^2
3) 1413 in^2
4) 401.92 yd^2

Test Preparation Answers

1) The answer is 8.

Use formula of rectangle prism volume.

V = (length) (width) (height) ⇒ 2000 = (25) (10) (height) ⇒

height = 2000 ÷ 250 = 8

2) Choice D is correct.

Use volume of a cube formula.

$V= a^3 \Rightarrow V= 4^3 \Rightarrow V= 64$

3) Choice D is correct.

Use volume of a cylinder formula.

$V= \pi r^2h \Rightarrow V= \pi (3)^2$ in × 8in ⇒ $V=72 \pi$ in^2

4) Choice D is correct

Volume of a box = length × width × height = 4 × 5 × 6 = 120

Chapter 12: Statistics

Topics that you'll learn in this chapter:

- ✓ Mean, Median, Mode, and Range of the Given Data
- ✓ The Pie Graph or Circle Graph
- ✓ Probability

Mean, Median, Mode, and Range of the Given Data

Helpful Hints

- Mean: $\frac{\text{sum of the data}}{\text{of data entires}}$
- Mode: value in the list that appears most often
- Range: largest value – smallest value

Example:

22, 16, 12, 9, 7, 6, 4, 6

Mean = 10.25

Mod = 6

Range = 18

Find Mean, Median, Mode, and Range of the Given Data.

1) 7, 2, 5, 1, 1, 2

2) 2, 2, 2, 3, 6, 3, 7, 4

3) 9, 4, 3, 1, 7, 9, 4, 6, 4

4) 8, 4, 2, 4, 3, 2, 4, 5

5) 8, 5, 7, 5, 7, 9, 8

6) 5, 1, 4, 4, 9, 2, 9, 2, 5, 1

7) 4, 1, 5, 9, 7, 7, 5, 4, 3, 5

8) 7, 5, 4, 9, 6, 7, 7, 5, 2

9) 2, 5, 5, 6, 2, 4, 7, 6, 4, 9

10) 10, 5, 2, 5, 4, 5, 8, 10

11) 5, 1, 5, 2, 2

12) 2, 3, 5, 9, 6

The Pie Graph or Circle Graph

Helpful Hints

A Pie Chart is a circle chart divided into sectors, each sector represents the relative size of each value.

The circle graph below shows all Jason's expenses for last month. Jason spent $300 on his bills last month.

Answer following questions based on the Pie graph.

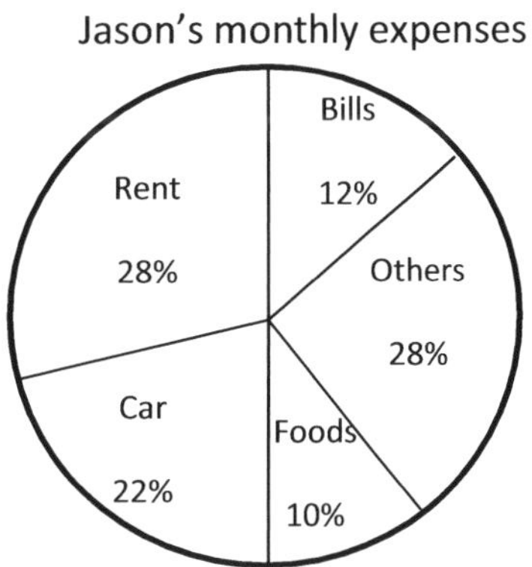

1- How much did Jason spend on his car last month?
2- How much did Jason spend for foods last month?
3- How much did Jason spend on his rent last month?
4- What fraction is Jason's expenses for his bills and Car out of his total expenses last month?
5- How much was Jason's senses last month?

Probability of Simple Events

Helpful Hints	- Probability is the likelihood of something happening in the future. It is expressed as a number between zero (can never happen) to 1 (will always happen). - Probability can be expressed as a fraction, a decimal, or a percent.	**Example:** Probability of a flipped coins turns up 'heads' Is $0.5 = \frac{1}{2}$

Solve.

1) A number is chosen at random from 1 to 10. Find the probability of selecting a 4 or smaller.

2) There are 135 blue balls and 15 red balls in a basket. What is the probability of randomly choosing a red ball from the basket?

3) A number is chosen at random from 1 to 10. Find the probability of selecting of 4 and factors of 6.

4) What is the probability of choosing a Hearts in a deck of cards? (A deck of cards contains 52 cards)

5) A number is chosen at random from 1 to 50. Find the probability of selecting prime numbers.

6) A number is chosen at random from 1 to 25. Find the probability of not selecting a composite number.

Test Preparation

1) Anita's trick–or–treat bag contains 12 pieces of chocolate, 18 suckers, 18 pieces of gum, 24 pieces of licorice. If she randomly pulls a piece of candy from her bag, what is the probability of her pulling out a piece of sucker?

A. $\frac{1}{3}$

B. $\frac{1}{4}$

C. $\frac{1}{6}$

D. $\frac{1}{12}$

E. $\frac{1}{20}$

2) The average of 6 numbers is 12. The average of 4 of those numbers is 10. What is the average of the other two numbers?

A. 10

B. 12

C. 14

D. 16

E. 20

3) The average of 13, 15, 20 and x is 18. What is the value of x?

Write your answer in the box below.

4) What is the median of these numbers? 4, 9, 13, 8, 15, 18, 5
   A. 8
   B. 9
   C. 13
   D. 15
   E. 20

5) The mean of 50 test scores was calculated as 88. But, it turned out that one of the scores was misread as 94 but it was 69. What is the mean?
   A. 85
   B. 87
   C. 87.5
   D. 88.5
   E. 90.5

6) Two dice are thrown simultaneously, what is the probability of getting a sum of 6 or 9?

A. $\frac{1}{3}$

B. $\frac{1}{4}$

C. $\frac{1}{6}$

D. $\frac{1}{12}$

E. $\frac{1}{22}$

7) What is the median of these numbers? 2, 27, 28, 19, 67, 44, 35

A. 19

B. 28

C. 44

D. 35

E. 40

8) The average of five numbers is 24. If a sixth number 42 is added, then, what is the new average?

A. 25

B. 26

C. 27

D. 28

E. 32

Answers of Worksheets – Chapter 12

Mean, Median, Mode, and Range of the Given Data

1) mean: 3, median: 2, mode: 1, 2, range: 6
2) mean: 3.625, median: 3, mode: 2, range: 5
3) mean: 5.22, median: 4, mode: 4, range: 8
4) mean: 4, median: 4, mode: 4, range: 6
5) mean: 7, median: 7, mode: 5, 7, 8, range: 4
6) mean: 4.2, median: 4, mode: 1,2,4,5,9, range: 8
7) mean: 5, median: 5, mode: 5, range: 8
8) mean: 5.78, median: 6, mode: 7, range: 7
9) mean: 5, median: 5, mode: 2, 4, 5, 6, range: 7
10) mean: 6.125, median: 5, mode: 5, range: 8
11) mean: 3, median: 2, mode: 2, 5, range: 4
12) mean: 5, median: 5, mode: none, range: 7

The Pie Graph or Circle Graph

1) $550
2) $250
3) $700
4) $\frac{17}{50}$
5) $2500

Probability of simple events

1) $\frac{2}{5}$
2) $\frac{1}{10}$
3) $\frac{1}{5}$
4) $\frac{1}{4}$
5) $\frac{3}{10}$
6) $\frac{2}{5}$

Test Preparation Answers

1) Choice B is correct.

Probability = $\frac{number\ of\ desired\ outcomes}{\text{number of total outcomes}} = \frac{18}{12+18+18+24} = \frac{18}{72} = \frac{1}{4}$

2) Choice D is correct

average = $\frac{\text{sum of terms}}{\text{number of terms}}$ ⇒ (average of 6 numbers) 12 = $\frac{\text{sum of numbers}}{6}$ ⇒sum of 6 numbers is 12 × 6 = 72

(average of 4 numbers) 10 = $\frac{\text{sum of numbers}}{4}$ ⇒sum of 4 numbers is 10 × 4 = 40

sum of 6 numbers – sum of 4 numbers = sum of 2 numbers

72 – 40 = 32, average of 2 numbers = $\frac{32}{2}$ = 16

3) The answer is 24.

average = $\frac{\text{sum of terms}}{\text{number of terms}}$ ⇒ (average of 4 numbers) 18 = $\frac{\text{sum of numbers}}{4}$ ⇒sum of 4 numbers is : 13+ 15+ 20 + x

72 = sum of numbers ⇒ 72= 13+ 15+ 20 + x ⇒ x=24

4) Choice B is correct

Write the numbers in order:

4, 5, 8, 9, 13, 15, 18

Since we have 7 numbers (7 is odd), then the median is the number in the middle, which is 9.

5) Choice C is correct

average (mean) = $\frac{\text{sum of terms}}{\text{number of terms}}$ ⇒ 88 = $\frac{\text{sum of terms}}{50}$ ⇒ sum = 88 × 50 = 4,400

The difference of 94 and 69 is 25. Therefore, 25 should be subtracted from the sum.

4400 – 25 = 4375

mean = $\frac{\text{sum of terms}}{\text{number of terms}}$ ⇒ mean = $\frac{4375}{50}$ = 87.5

6) Choice B is correct

For Sum 6: (1 & 5) and (5 & 1), (2 & 4) and (4 & 2), (3 & 3), so we have 5 options

For sum 9: (3 & 6) and (6 & 3), (4 & 5) and (5 & 4), we have 4 options.

To get a sum of 6 or 9 for two dice: 7+4=9

Since, we have 6 × 6 = 36 total options, the probability of getting a sum of 6 and 9 is 11 out of 36 or $\frac{9}{36} = \frac{1}{4}$

7) Choice B is correct

Write the numbers in order:

2, 19, 27, 28, 35, 44, 67

Median is the number in the middle. So, the median is 28.

8) Choice C is correct

Solve for the sum of five numbers.

average = $\frac{\text{sum of terms}}{\text{number of terms}}$ ⇒ 24 = $\frac{\text{sum of 5 numbers}}{5}$ ⇒ sum of 5 numbers = 24 × 5 = 120

The sum of 5 numbers is 120. If a sixth number 42 is added, then the sum of 6 numbers is

120 + 42 = 162

average = $\frac{\text{sum of terms}}{\text{number of terms}}$ = $\frac{162}{6}$ = 27

SSAT Middle Level Mathematics Practice Tests

The SSAT, or Secondary School Admissions Test, is a standardized test to help determine admission to private elementary, middle and high schools.

There are currently three Levels of the SSAT:

- ✓ Lower Level (for students in 3rd and 4th grade)
- ✓ Middle Level (for students in 5th-7th grade)
- ✓ Upper Level (for students in 8th-11th grade)

There are six sections on the SSAT Middle Level Test:

- ✓ Writing: 25 minutes.
- ✓ Math section: 25 questions, 30 minutes
- ✓ Reading section: 40 questions, 40 minutes
- ✓ Verbal section: 60 questions, 30 minutes
- ✓ Math section: 25 questions, 30 minutes
- ✓ Experimental: 16 questions, 15 minutes.

In this book, we have reviewed mathematics topics being tested on the math sections of the SSAT Middle Level. In this section, there are two complete SSAT Middle Level Math Tests. Let your student take these tests to see what score they will be able to receive on a real SSAT Middle Level test.

Good luck!

Time to Test

Time to refine your skill with a practice examination

Take two practice SSAT Middle Level Mathematics Tests to simulate the test day experience. After you've finished, score your test using the answer keys.

Before You Start

- You'll need a pencil and a timer to take the test.
- After you've finished the test, review the answer key to see where you went wrong.
- Use the answer sheet provided to record your answers. (You can cut it out or photocopy it)
- You will receive 1 point for every correct answer and you will lose $\frac{1}{4}$ point for each incorrect answer. There is no penalty for skipping a question.

Calculators are NOT permitted for the SSAT Middle Level Test

Good Luck!

SSAT Practice Test 1 Answer Sheet

Remove (or photocopy) this answer sheet and use it to complete the practice test.

SSAT Middle Level Mathematics Practice Test 1 Answer Sheet

SSAT Middle Level Practice Test 1 Section 1

1	Ⓐ Ⓑ Ⓒ Ⓓ Ⓔ	11	Ⓐ Ⓑ Ⓒ Ⓓ Ⓔ	21	Ⓐ Ⓑ Ⓒ Ⓓ Ⓔ
2	Ⓐ Ⓑ Ⓒ Ⓓ Ⓔ	12	Ⓐ Ⓑ Ⓒ Ⓓ Ⓔ	22	Ⓐ Ⓑ Ⓒ Ⓓ Ⓔ
3	Ⓐ Ⓑ Ⓒ Ⓓ Ⓔ	13	Ⓐ Ⓑ Ⓒ Ⓓ Ⓔ	23	Ⓐ Ⓑ Ⓒ Ⓓ Ⓔ
4	Ⓐ Ⓑ Ⓒ Ⓓ Ⓔ	14	Ⓐ Ⓑ Ⓒ Ⓓ Ⓔ	24	Ⓐ Ⓑ Ⓒ Ⓓ Ⓔ
5	Ⓐ Ⓑ Ⓒ Ⓓ Ⓔ	15	Ⓐ Ⓑ Ⓒ Ⓓ Ⓔ	25	Ⓐ Ⓑ Ⓒ Ⓓ Ⓔ
6	Ⓐ Ⓑ Ⓒ Ⓓ Ⓔ	16	Ⓐ Ⓑ Ⓒ Ⓓ Ⓔ		
7	Ⓐ Ⓑ Ⓒ Ⓓ Ⓔ	17	Ⓐ Ⓑ Ⓒ Ⓓ Ⓔ		
8	Ⓐ Ⓑ Ⓒ Ⓓ Ⓔ	18	Ⓐ Ⓑ Ⓒ Ⓓ Ⓔ		
9	Ⓐ Ⓑ Ⓒ Ⓓ Ⓔ	19	Ⓐ Ⓑ Ⓒ Ⓓ Ⓔ		
10	Ⓐ Ⓑ Ⓒ Ⓓ Ⓔ	20	Ⓐ Ⓑ Ⓒ Ⓓ Ⓔ		

SSAT Middle Level Practice Test 1 Section 2

1	Ⓐ Ⓑ Ⓒ Ⓓ Ⓔ	11	Ⓐ Ⓑ Ⓒ Ⓓ Ⓔ	21	Ⓐ Ⓑ Ⓒ Ⓓ Ⓔ
2	Ⓐ Ⓑ Ⓒ Ⓓ Ⓔ	12	Ⓐ Ⓑ Ⓒ Ⓓ Ⓔ	22	Ⓐ Ⓑ Ⓒ Ⓓ Ⓔ
3	Ⓐ Ⓑ Ⓒ Ⓓ Ⓔ	13	Ⓐ Ⓑ Ⓒ Ⓓ Ⓔ	23	Ⓐ Ⓑ Ⓒ Ⓓ Ⓔ
4	Ⓐ Ⓑ Ⓒ Ⓓ Ⓔ	14	Ⓐ Ⓑ Ⓒ Ⓓ Ⓔ	24	Ⓐ Ⓑ Ⓒ Ⓓ Ⓔ
5	Ⓐ Ⓑ Ⓒ Ⓓ Ⓔ	15	Ⓐ Ⓑ Ⓒ Ⓓ Ⓔ	25	Ⓐ Ⓑ Ⓒ Ⓓ Ⓔ
6	Ⓐ Ⓑ Ⓒ Ⓓ Ⓔ	16	Ⓐ Ⓑ Ⓒ Ⓓ Ⓔ		
7	Ⓐ Ⓑ Ⓒ Ⓓ Ⓔ	17	Ⓐ Ⓑ Ⓒ Ⓓ Ⓔ		
8	Ⓐ Ⓑ Ⓒ Ⓓ Ⓔ	18	Ⓐ Ⓑ Ⓒ Ⓓ Ⓔ		
9	Ⓐ Ⓑ Ⓒ Ⓓ Ⓔ	19	Ⓐ Ⓑ Ⓒ Ⓓ Ⓔ		
10	Ⓐ Ⓑ Ⓒ Ⓓ Ⓔ	20	Ⓐ Ⓑ Ⓒ Ⓓ Ⓔ		

SSAT Middle Level

Mathematics Practice Test 1

Section 1

25 questions

Total time for this test: 30 Minutes

You may NOT use a calculator for this test.

1) If 15 percent of a number is 60, then 25 percent of the same number is …

(A) 65

(B) 70

(C) 80

(D) 100

(E) 120

---

2) Which of the following is NOT equal to 0.2×4?

(A) 0.4×2

(B) 1×0.8

(C) $\frac{16}{8} \times \frac{4}{10}$

(D) $\frac{5}{15} \times 3$

(E) 0.8×1

---

3) Sara has N books. Mary has 5 more books than Sara. If Mary gives Sara 4 books, how many books will Mary have, in terms of N?

(A) N

(B) $N + 1$

(C) $N + 2$

(D) $N + 5$

(E) $N - 5$

---

4) If $\frac{3x}{2} = 30$, then $\frac{2x}{5}$=?

(A) 8

(B) 10

(C) 15

(D) 20

(E) 40

---

5) Which of the following is closest to $\frac{1}{5}$ of 40?

(A) 0.3×6

(B) 0.3×5

(C) 0.2×30

(D) 0.2×35

(E) 0.2×39.5

---

6) What is the area of a square whose diagonal is 6?

(A) 18

(B) 24

(C) 36

(D) 60

(E) 64

7) An angle is equal to one eighth of its supplement. What is the measure of that angle?

(A) 15

(B) 20

(C) 30

(D) 45

(E) 160

---

8) A $40 shirt now selling for $28 is discounted by what percent?

(A) 20 %

(B) 30 %

(C) 40 %

(D) 60 %

(E) 80 %

---

9) What is the value of x in the following figure?

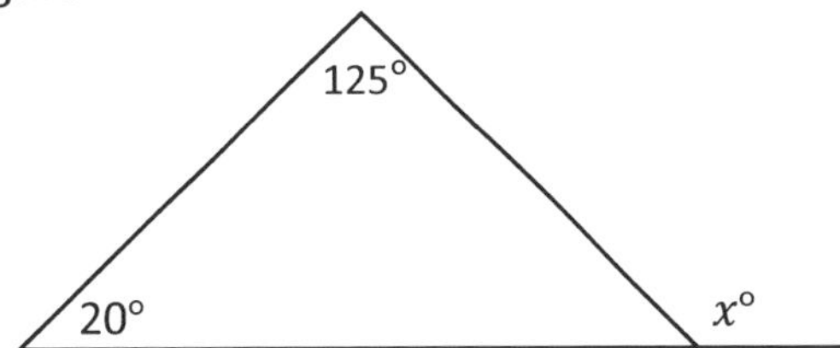

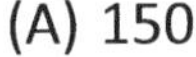
(A) 150

(B) 145

(C) 125

(D) 105

(E) 85

10) The perimeter of the trapezoid below is 54. What is its area?

(A) 252 cm^2

(B) 234 cm^2

(C) 216 cm^2

(D) 130 cm^2

(E) 108 cm^2

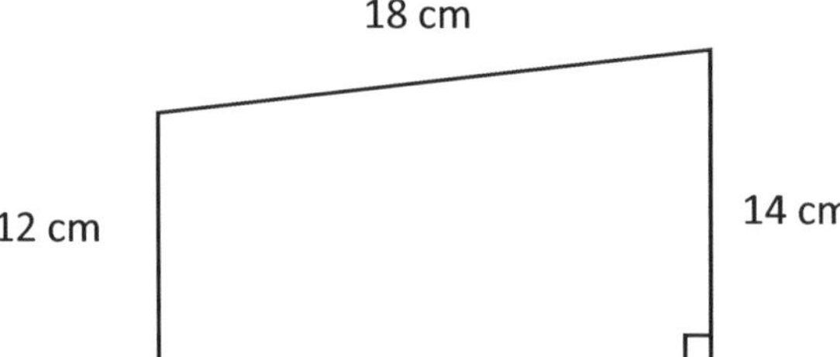

---

11) The score of Emma was half as that of Ava and the score of Mia was twice that of Ava. If the score of Mia was 60, what is the score of Emma?

(A) 15

(B) 18

(C) 20

(D) 30

(E) 32

---

12) Two third of 18 is equal to $\frac{2}{5}$ of what number?

(A) 60

(B) 20

(C) 12

(D) 30

(E) 36

13) If three times a number added to 6 equals to 30, what is the number?

(A) 2

(B) 4

(C) 6

(D) 8

(E) 10

---

14) Solve for x: $4(x + 1) = 6(x - 4) + 20$

(A) 12

(B) 6.2

(C) 5.5

(D) 4

(E) 2

---

15) Five years ago, Amy was three times as old as Mike was. If Mike is 10 years old now, how old is Amy?

(A) 4

(B) 8

(C) 12

(D) 15

(E) 20

16) Two-kilograms apple and three-kilograms orange cost \$26.4. If one-kilogram apple costs \$4.2 how much does one-kilogram orange cost?

(A) \$9

(B) \$6

(C) \$5.5

(D) \$5

(E) \$4.5

---

17) The average weight of 18 girls in a class is 60 kg and the average weight of 32 boys in the same class is 62 kg. What is the average weight of all the 50 students in that class?

(A) 61.28

(B) 61.68

(C) 61.90

(D) 62.20

(E) 64.00

---

18) What is the value of x in this equation? $6(x + 4) = 72$

(A) 4

(B) 6

(C) 8

(D) 10

(E) 12

19) When a number is subtracted from 24 and the difference is divided by that number, the result is 3. What is the value of the number?

(A) 2

(B) 4

(C) 6

(D) 12

(E) 15

---

20) Which of the following is the correct statement?

(A) $\frac{3}{4} > 0.8$

(B) $10\% = \frac{2}{5}$

(C) $3 < \frac{5}{2}$

(D) $\frac{5}{6} > 0.8$

(E) $2.5\% = 0.25$

---

21) In a group of 5 books, the average number of pages is 24. Mary adds a book with 30 pages to the group. What is the new average number of pages per book?

(A) 20

(B) 22

(C) 24

(D) 25

(E) 30

22) A football team won exactly 80% of the games it played during last session. Which of the following could be the total number of games the team played last season?

(A) 49

(B) 35

(C) 32

(D) 12

(E) 8

---

23) If a gas tank can hold 25 gallons, how many gallons does it contain when it is $\frac{2}{5}$ full?

(A) 50

(B) 125

(C) 62.5

(D) 10

(E) 8

---

24) A red box is 20% greater than the blue box. If 30 books exist in the red box, how many books are in the blue box?

(A) 9

(B) 15

(C) 20

(D) 25

(E) 26

25) 6 liters of water are poured into an aquarium that's 15 cm long, 5 cm wide, and 60 cm high. How many cm will the water level in the aquarium rise due to this added water? (1 liter of water = 1,000 cm3)

(A) 80

(B) 40

(C) 20

(D) 10

(E) 8

IF YOU FINISH BEFORE TIME IS CALLED, YOU MAY CHECK YOUR WORK ON THIS SECTION ONLY. DO NOT TURN TO OTHER SECTION IN THE TEST.	STOP

SSAT Middle Level

Mathematics Practice Test 1

Section 2

25 questions

Total time for this test: 30 Minutes

You may NOT use a calculator for this test.

1) A taxi driver earns $9 per 1-hour work. If he works 10 hours a day and in 1 hour he uses 2-liters petrol with price $1 for 1-liter. How much money does he earn in one day?

(A) $90

(B) $88

(C) $70

(D) $60

(E) $56

---

2) Which of the following is less than $\frac{1}{5}$?

(A) $\frac{1}{4}$

(B) 0.5

(C) $\frac{1}{6}$

(D) 0.25

(E) 0.3

---

3) Amy and John work in a same company. Last month, both of them received a raise of 20 percent. If Amy earns $30.00 per hour now and John earns $26.40, Amy earned how much more per hour than John before their raises?

(A) $8.25

(B) $4.25

(C) $3.00

(D) $2.25

(E) $1.75

---

4) Four people can paint 4 houses in 10 days. How many people are needed to paint 8 houses in 5 days?

(A) 6

(B) 8

(C) 12

(D) 16

(E) 20

---

5) If $N \times (5 - 3) = 12$ then $N = ?$

(A) 6

(B) 12

(C) 13

(D) 14

(E) 18

---

6) The length of a rectangle is 3 times of its width. If the length is 18, what is the perimeter of the rectangle?

(A) 24

(B) 30

(C) 36

(D) 48

(E) 56

7) In the figure below, what is the value of x?

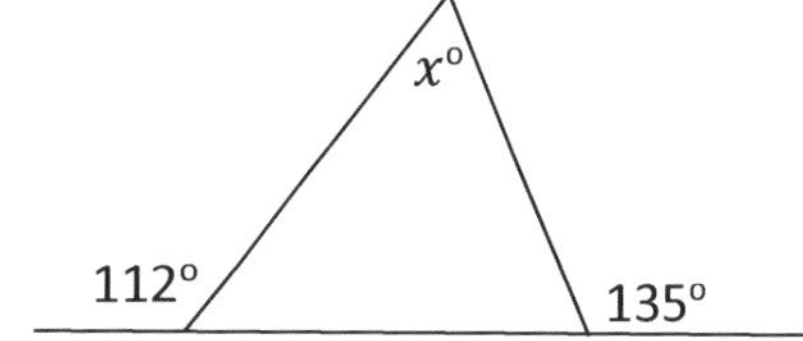

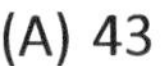
(A) 43

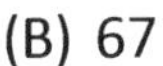
(B) 67

(C) 77

(D) 90

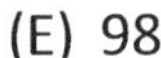
(E) 98

---

8) If $x \blacksquare y = 3x + y - 2$, what is the value of $4 \blacksquare 12$?

(A) 4

(B) 18

(C) 22

(D) 36

(E) 48

---

9) The width of a rectangle is $4x$. the length is $6x$, and the perimeter of the rectangle is 80. What is the value of x?

(A) 1

(B) 2

(C) 3

(D) 4

(E) 5

10) How many tiles of 8 cm^2 is needed to cover a floor of dimension 6 cm by 24 cm?

(A) 6

(B) 12

(C) 18

(D) 24

(E) 30

---

11) If 0.45 equals $45M$, what is the value of M?

(A) 0.10

(B) 0.01

(C) 1.00

(D) 11.01

(E) 0.11

---

12) If $z = 3x + 6$, what does $2z + 3$ equal?

(A) $6x + 6$

(B) $6x + 12$

(C) $6x - 12$

(D) $6x - 6$

(E) $6x + 15$

13) If 20 can be divided by both 4 and x without leaving a remainder, then 20 can also be divided by which of the following?

(A) $x + 4$

(B) $2x - 4$

(C) $x - 2$

(D) $x \times 4$

(E) $x + 1$

0.ABC 0.0D

14) The letters represent two decimals listed above. One of the decimals is equivalent to $\frac{1}{8}$ and the other is equivalent to $\frac{1}{20}$. What is the product of C and D?

(A) 0

(B) 5

(C) 25

(D) 20

(E) 40

---

15) $\frac{x}{x-3} = \frac{4}{5}, x - 5 = ?$

(A) -12

(B) -15

(C) -17

(D) 12

(E) 15

16) A company pays its employee \$7,000 plus 2% of all sales profit. If x is the number of all sales profit, which of the following represents the employer's revenue?

(A) $0.02x$

(B) $0.98x - 7{,}000$

(C) $0.02x + 7{,}000$

(D) $0.98x + 7{,}000$

(E) $0.2x + 7{,}000$

---

17) In a certain bookshelf of a library, there are 35 biology books, 95 history books, and 80 language books. What is the ratio of the number of biology books to the total number of books in this bookshelf?

(A) $\frac{1}{4}$

(B) $\frac{1}{6}$

(C) $\frac{2}{7}$

(D) $\frac{3}{8}$

(E) $\frac{1}{4}$

---

18) If $6{,}000 + A - 200 = 7{,}400$, then $A =$

(A) 200

(B) 600

(C) 1,600

(D) 2,200

(E) 3,000

19) The circle graph below shows all Mr. Green's expenses for last month. If he spent $660 on his car, how much did he spend for his rent?

Mr. Green's monthly expenses

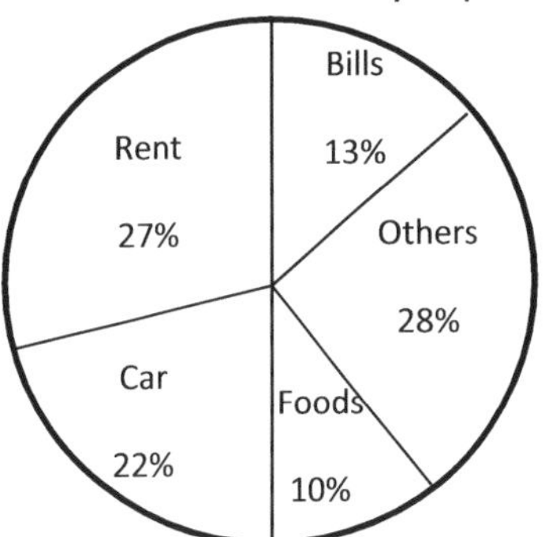

(A) $700

(B) $740

(C) $780

(D) $810

(E) $900

---

20) If $5 \times M + 4 = 5$, M equals to ....

(A) 2

(B) 4

(C) $\frac{1}{5}$

(D) 6

(E) $\frac{1}{3}$

---

21) Which of the following is equal to $\frac{42.6}{100}$?

(A) 42.6

(B) 4.26

(C) 426.0

(D) 0.0426

(E) 0.426

22) In the following figure, point Q lies on line n, what is the value of y if $x = 35$?

(A) 15

(B) 25

(C) 35

(D) 45

(E) 60

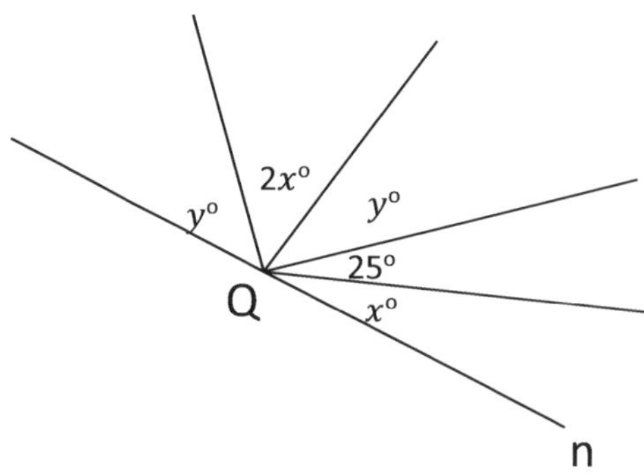

23) A container holds 3.5 gallons of water when it is $\frac{7}{24}$ full. How many gallons of water does the container hold when it's full?

(A) 8

(B) 12

(C) 16

(D) 20

(E) 30

24) At a Zoo, the ratio of lions to tigers is 5 to 3. Which of the following could NOT be the total number of lions and tigers in the zoo?

(A) 64

(B) 80

(C) 98

(D) 104

(E) 160

25) If x is greater than 18, then $\frac{1}{6}$ of x must be...

(A) Greater than 3

(B) Greater than 6

(C) Equal to 6

(D) Equal to 3

(E) Less than 3

IF YOU FINISH BEFORE TIME IS CALLED, YOU MAY CHECK YOUR WORK ON THIS SECTION ONLY. DO NOT TURN TO ANY OTHER SECTION IN THE TEST. STOP

SSAT Practice Test 2 Answer Sheet

Remove (or photocopy) this answer sheet and use it to complete the practice test.

SSAT Middle Level Mathematics Practice Test 2 Answer Sheet

SSAT Middle Level Practice Test 2 Section 1

1	Ⓐ Ⓑ Ⓒ Ⓓ Ⓔ	11	Ⓐ Ⓑ Ⓒ Ⓓ Ⓔ	21	Ⓐ Ⓑ Ⓒ Ⓓ Ⓔ
2	Ⓐ Ⓑ Ⓒ Ⓓ Ⓔ	12	Ⓐ Ⓑ Ⓒ Ⓓ Ⓔ	22	Ⓐ Ⓑ Ⓒ Ⓓ Ⓔ
3	Ⓐ Ⓑ Ⓒ Ⓓ Ⓔ	13	Ⓐ Ⓑ Ⓒ Ⓓ Ⓔ	23	Ⓐ Ⓑ Ⓒ Ⓓ Ⓔ
4	Ⓐ Ⓑ Ⓒ Ⓓ Ⓔ	14	Ⓐ Ⓑ Ⓒ Ⓓ Ⓔ	24	Ⓐ Ⓑ Ⓒ Ⓓ Ⓔ
5	Ⓐ Ⓑ Ⓒ Ⓓ Ⓔ	15	Ⓐ Ⓑ Ⓒ Ⓓ Ⓔ	25	Ⓐ Ⓑ Ⓒ Ⓓ Ⓔ
6	Ⓐ Ⓑ Ⓒ Ⓓ Ⓔ	16	Ⓐ Ⓑ Ⓒ Ⓓ Ⓔ		
7	Ⓐ Ⓑ Ⓒ Ⓓ Ⓔ	17	Ⓐ Ⓑ Ⓒ Ⓓ Ⓔ		
8	Ⓐ Ⓑ Ⓒ Ⓓ Ⓔ	18	Ⓐ Ⓑ Ⓒ Ⓓ Ⓔ		
9	Ⓐ Ⓑ Ⓒ Ⓓ Ⓔ	19	Ⓐ Ⓑ Ⓒ Ⓓ Ⓔ		
10	Ⓐ Ⓑ Ⓒ Ⓓ Ⓔ	20	Ⓐ Ⓑ Ⓒ Ⓓ Ⓔ		

SSAT Middle Level Practice Test 2 Section 2

1	Ⓐ Ⓑ Ⓒ Ⓓ Ⓔ	11	Ⓐ Ⓑ Ⓒ Ⓓ Ⓔ	21	Ⓐ Ⓑ Ⓒ Ⓓ Ⓔ
2	Ⓐ Ⓑ Ⓒ Ⓓ Ⓔ	12	Ⓐ Ⓑ Ⓒ Ⓓ Ⓔ	22	Ⓐ Ⓑ Ⓒ Ⓓ Ⓔ
3	Ⓐ Ⓑ Ⓒ Ⓓ Ⓔ	13	Ⓐ Ⓑ Ⓒ Ⓓ Ⓔ	23	Ⓐ Ⓑ Ⓒ Ⓓ Ⓔ
4	Ⓐ Ⓑ Ⓒ Ⓓ Ⓔ	14	Ⓐ Ⓑ Ⓒ Ⓓ Ⓔ	24	Ⓐ Ⓑ Ⓒ Ⓓ Ⓔ
5	Ⓐ Ⓑ Ⓒ Ⓓ Ⓔ	15	Ⓐ Ⓑ Ⓒ Ⓓ Ⓔ	25	Ⓐ Ⓑ Ⓒ Ⓓ Ⓔ
6	Ⓐ Ⓑ Ⓒ Ⓓ Ⓔ	16	Ⓐ Ⓑ Ⓒ Ⓓ Ⓔ		
7	Ⓐ Ⓑ Ⓒ Ⓓ Ⓔ	17	Ⓐ Ⓑ Ⓒ Ⓓ Ⓔ		
8	Ⓐ Ⓑ Ⓒ Ⓓ Ⓔ	18	Ⓐ Ⓑ Ⓒ Ⓓ Ⓔ		
9	Ⓐ Ⓑ Ⓒ Ⓓ Ⓔ	19	Ⓐ Ⓑ Ⓒ Ⓓ Ⓔ		
10	Ⓐ Ⓑ Ⓒ Ⓓ Ⓔ	20	Ⓐ Ⓑ Ⓒ Ⓓ Ⓔ		

SSAT Middle Level

Mathematics Practice Test 2

Section 1

25 questions

Total time for this test: 30 Minutes

You may NOT use a calculator for this test.

1. How long does a 420–miles trip take moving at 50 miles per hour (mph)?
   (A) 4 hours
   (B) 6 hours and 24 minutes
   (C) 8 hours and 24 minutes
   (D) 8 hours and 30 minutes
   (E) 10 hours and 30 minutes

---

2. The marked price of a computer is D dollar. Its price decreased by 20% in January and later increased by 10 % in February. What is the final price of the computer in D dollar?
   (A) 0.80 D
   (B) 0.88 D
   (C) 0.90 D
   (D) 1.20 D
   (E) 1.40 D

---

3. If 0.55 equals 55M, what is the value of M?
   (A) 0.01
   (B) 0.1
   (C) 1.0
   (D) 1.01
   (E) 1.001

4. Jason borrowed \$4,800 for three months at an annual rate of 5%. How much interest did Jason owe?
   (A) \$45
   (B) \$60
   (C) \$120
   (D) \$240
   (E) \$480

---

5. If three times a certain number, increased by 12, is equal to 39, what is the number?
   (A) 9
   (B) 12
   (C) 18
   (D) 27
   (E) 54

---

6. If 30 percent of a number is 150, then 12 percent of the same number is
   (A) 60
   (B) 72
   (C) 80
   (D) 90
   (E) 120

---

7. The average of 13, 15, 20 and x is 18. What is the value of x?

   (A) 9

   (B) 15

   (C) 18

   (D) 20

   (E) 24

---

8. In five successive hours, a car travels 40 km, 45 km, 50 km, 35 km and 55 km. In the next five hours, it travels with an average speed of 50 km per hour. Find the total distance the car traveled in 10 hours.

   (A) 425 km

   (B) 450 km

   (C) 475 km

   (D) 500 km

   (E) 1,000 km

---

9. John has N toy cars. Jack has 4 more cars than John. If Jack gives John 3 cars, how many cars will Jack have, in terms of N?

   (A) N

   (B) N - 1

   (C) N + 1

   (D) N + 2

   (E) N + 3

10. What is the value of x in the following equation?

$$\frac{x+4}{5} = 2$$

(A) 2

(B) 4

(C) 6

(D) 8

(E) 10

---

11. The ratio of boys to girls in a school is 2:3. If there are 600 students in a school, how many boys are in the school.

(A) 540

(B) 360

(C) 300

(D) 280

(E) 240

---

12. Two third of 18 is equal to $\frac{2}{5}$ of what number?

(A) 12

(B) 20

(C) 30

(D) 60

(E) 90

13. What is the cost of six ounces of cheese at $0.96 per pound?

(A) $0.36

(B) $0.40

(C) $0.48

(D) $0.52

(E) $0.64

---

14. If 60 % of A is 20 % of B, then B is what percent of A?

(A) 3 %

(B) 30 %

(C) 200 %

(D) 300 %

(E) 900 %

---

15. Sophia purchased a sofa for $530.40. The sofa is regularly priced at $624. What was the percent discount Sophia received on the sofa?

(A) 12%

(B) 15%

(C) 20%

(D) 25%

(E) 40%

16. A bag contains 18 balls: two green, five black, eight blue, a brown, a red and one white. If 17 balls are removed from the bag at random, what is the probability that a brown ball has been removed?

(A) $\frac{1}{9}$

(B) $\frac{1}{6}$

(C) $\frac{16}{17}$

(D) $\frac{17}{18}$

(E) $\frac{1}{2}$

---

17. When a number is subtracted from 24 and the difference is divided by that number, the result is 3. What is the value of the number?

(A) 2

(B) 4

(C) 6

(D) 12

(E) 24

---

18. If 40 % of a class are girls, and 25 % of girls play tennis, what percent of the class play tennis?

(A) 10 %

(B) 15%

(C) 20 %

(D) 40 %

(E) 80 %

19. 55 students took an exam and 11 of them failed. What percent of the students passed the exam?

(A) 20 %

(B) 40 %

(C) 60 %

(D) 80 %

(E) 90 %

---

20. What is the value of x in the following equation?

$$3x + 10 = 49$$

(A) 5

(B) 7

(C) 9

(D) 11

(E) 13

---

21. If $N \times \frac{4}{3} \times 5 = 0$, then $N =$....

(A) 0

(B) 1

(C) 2

(D) 3

(E) 4

22. Jason left a $12.00 tip on a lunch that cost $48.00, approximately what percentage was the tip?

(A) 2.5%

(B) 10%

(C) 15%

(D) 20%

(E) 25%

---

23. If 40 % of a number is 4, what is the number?

(A) 4

(B) 8

(C) 10

(D) 12

(E) 20

---

24. If $\frac{z}{5} = 5$, then $z + 3 =?$

(A) 4

(B) 5

(C) 15

(D) 20

(E) 28

25. In 1999, the average worker's income increased $2,000 per year starting from $24,000 annual salary. Which equation represents income greater than average? (I = income, x = number of years after 1999)

(A) $I > 2000x + 24000$

(B) $I >-2000x + 24000$

(C) $I <-2000x + 24000$

(D) $I < 2000x - 24000$

(E) $I < 24{,}000x + 24000$

IF YOU FINISH BEFORE TIME IS CALLED, YOU MAY CHECK YOUR WORK ON THIS SECTION ONLY. DO NOT TURN TO ANY OTHER SECTION IN THE TEST.	STOP

SSAT Middle Level

Mathematics Practice Test 2

Section 2

25 questions

Total time for this test: 30 Minutes

You may NOT use a calculator for this test.

1. John has x dollars and he receives \$150. He then buys a bicycle that costs \$120. How much money does John have now?

   (A) $x + 150$

   (B) $x + 120$

   (C) $x + 30$

   (D) $x - 120$

   (E) $x - 30$

---

2. What is the value of x in this equation?

$$\frac{x-3}{8} + 4 = 20$$

   (A) 131

   (B) 128

   (C) 124

   (D) 120

   (E) 115

---

3. Jason needs an 75% average in his writing class to pass. On his first 4 exams, he earned scores of 68%, 72%, 85%, and 90%. What is the minimum score Jason can earn on his fifth and final test to pass?

   (A) 80%,

   (B) 70%

   (C) 68%

   (D) 64%

   (E) 60%

---

4. The width of a rectangle is $6x$, the length is $8x$, and the perimeter is 56. What is the value of x?

(A) 1

(B) 2

(C) 3

(D) 4

(E) 5

---

5. A bank is offering 4.5% simple interest on a savings account. If you deposit $8,000, how much interest will you earn in five years?

(A) $360

(B) $720

(C) $1,800

(D) $3,600

(E) $4,800

---

6. If $(9 - 5) \times 4 = 8 + \square$, then $\square$=?

(A) 5

(B) 6

(C) 7

(D) 8

(E) 9

7. Jason is 9 miles ahead of Joe running at 5.5 miles per hour and Joe is running at the speed of 7 miles per hour. How long does it take Joe to catch Jason?
   (A) 3 hours
   (B) 4 hours
   (C) 6 hours
   (D) 8 hours
   (E) 10 hours

8. In a classroom, there are y tables that can each seat 5 people and there are x tables that can each seat 8 people. What is the number of people that can be seated in the classroom?
   (A) $5y$
   (B) $8x$
   (C) $8x - 5y$
   (D) 13
   (E) $8x + 5y$

9. The area of a circle is 64π. What is the diameter of the circle?
   (A) 4
   (B) 8
   (C) 12
   (D) 14
   (E) 16

10. A shirt costing $200 is discounted 15%. After a month, the shirt is discounted another 15%. Which of the following expressions can be used to find the selling price of the shirt?

(A) (200) (0.70)

(B) (200) – 200 (0.30)

(C) (200) (0.15) – (200) (0.15)

(D) (200) (0.85) (0.85)

(E) (200) (0.85) (0.85) – (200) (0.15)

---

11. Four one – foot rulers can be split among how many users to leave each with $\frac{1}{6}$ of a ruler?

(A) 4

(B) 6

(C) 12

(D) 24

(E) 48

---

12. The perimeter of a rectangular yard is 60 meters. What is its length if its width is twice its length?

(A) 10 meters

(B) 18 meters

(C) 20 meters

(D) 24 meters

(E) 36 meters

---

13. What is the value of x in this equation? $3x + 12 = 48$

(A) 18

(B) 14

(C) 12

(D) 10

(E) 6

---

14. The mean of 50 test scores was calculated as 88. But, it turned out that one of the scores was misread as 94 but it was 69. What is the mean?

(A) 85

(B) 87

(C) 87.5

(D) 88.5

(E) 90.5

---

15. The average of 6 numbers is 12. The average of 4 of those numbers is 10. What is the average of the other two numbers?

(A) 10

(B) 12

(C) 14

(D) 16

(E) 24

---

16. If $x + 5 = 8$, $2y - 1 = 5$ then $xy + 10 =$

(A) 10

(B) 19

(C) 21

(D) 27

(E) 32

---

17. A card is drawn at random from a standard 52–card deck, what is the probability that the card is of Hearts? (The deck includes 13 of each suit clubs, diamonds, hearts, and spades)

(A) $\frac{1}{3}$

(B) $\frac{1}{4}$

(C) $\frac{1}{6}$

(D) $\frac{1}{52}$

(E) $\frac{1}{104}$

---

18. Which of the following is NOT less than $\frac{1}{5}$?

(A) $\frac{1}{8}$

(B) $\frac{1}{4}$

(C) $\frac{1}{9}$

(D) $\frac{1}{10}$

(E) $\frac{1}{7}$

19. Mr. Jones saves $2,500 out of his monthly family income of $55,000. What fractional part of his income does he save?

(A) $\frac{1}{22}$

(B) $\frac{1}{11}$

(C) $\frac{3}{25}$

(D) $\frac{2}{15}$

(E) $\frac{1}{15}$

---

20. If $x = 9$, then $2x + 6 =?$

(A) 18

(B) 20

(C) 22

(D) 24

(E) 26

---

21. In two successive years, the population of a town is increased by 15% and 20%. What percent of the population is increased after two years?

(A) 32%

(B) 35%

(C) 38%

(D) 68%

(E) 70%

22. If 150 % of a number is 75, then what is the 90 % of that number?

(A) 45

(B) 50

(C) 70

(D) 85

(E) 90

23. What is the equivalent temperature of 104°F in Celsius? (C = Celsius)

$$C = \frac{5}{9}(F - 32)$$

(A) 32

(B) 40

(C) 48

(D) 52

(E) 64

24. The perimeter of the trapezoid below is 36 cm. What is its area?

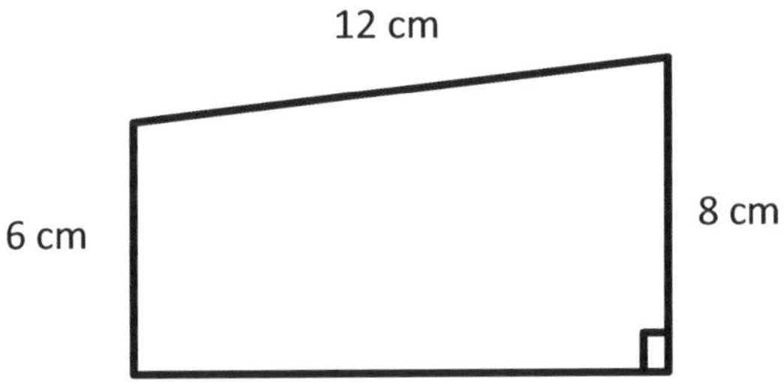

(A) 48 cm^2

(B) 70 cm^2

(C) 140 cm^2

(D) 576 cm^2

(E) 986 cm^2

---

25. The width of a box is one third of its length. The height of the box is one third of its width. If the length of the box is 27 cm, what is the volume of the box?

(A) 81 cm^3

(B) 162 cm^3

(C) 243 cm^3

(D) 729 cm^3

(E) 1880 cm^3

IF YOU FINISH BEFORE TIME IS CALLED, YOU MAY CHECK YOUR WORK ON THIS SECTION ONLY. DO NOT TURN TO OTHER SECTION IN THE TEST. **STOP**

SSAT Middle Level Mathematics Practice Tests Answers and Explanations

SSAT Middle Level Mathematics Practice Test 1

Section 1				Section 2			
1	D	16	B	1	C	16	C
2	D	17	A	2	C	17	B
3	B	18	C	3	C	18	C
4	A	19	C	4	D	19	D
5	E	20	D	5	A	20	C
6	A	21	D	6	D	21	E
7	B	22	B	7	B	22	B
8	B	23	D	8	C	23	B
9	B	24	D	9	D	24	C
10	D	25	A	10	C	25	A
11	A			11	B		
12	D			12	E		
13	D			13	D		
14	D			14	C		
15	E			15	C		

SSAT Middle Level Mathematics Practice Test 2

Section 1				Section 2			
1	**C**	16	**D**	1	**C**	16	**B**
2	**B**	17	**C**	2	**A**	17	**B**
3	**A**	18	**A**	3	**E**	18	**B**
4	**B**	19	**D**	4	**B**	19	**A**
5	**A**	20	**E**	5	**C**	20	**D**
6	**A**	21	**A**	6	**D**	21	**C**
7	**E**	22	**E**	7	**C**	22	**A**
8	**C**	23	**C**	8	**E**	23	**B**
9	**C**	24	**E**	9	**E**	24	**B**
10	**C**	25	**A**	10	**D**	25	**D**
11	**E**			11	**D**		
12	**C**			12	**A**		
13	**A**			13	**C**		
14	**D**			14	**C**		
15	**B**			15	**D**		

Score Your Test

SSAT scores are broken down by its three sections: Verbal, Mathematics, and Reading. A sum of the three sections is also reported.

For the Middle Level SSAT, the score range is 500-800, the lowest possible score a student can earn is 500 and the highest score is 800 for each section. A student receives 1 point for every correct answer and loses $\frac{1}{4}$ point for each incorrect answer. No points are lost by skipping a question.

The total scaled score for a Middle Level SSAT test is the sum of the scores for the Mathematics, verbal, and reading sections. A student will also receive a percentile score of between 1-99% that compares that student's test scores with those of other test takers of same grade and gender from the past 3 years.

Use the following table to convert SSAT Middle level raw score to scaled score.

SSAT Middle Level Math Scaled Scores	
Raw Scores	**Scaled Score**
50	710
45	680
40	660
35	635
30	615
25	590
20	570
15	540
10	525
5	500
0	480
-5	460
- 10 and lower	440

SSAT Middle Level Mathematics Practice Test 1

Section 1

1) Choice D is correct

If 15 percent of a number is 60, then the number is:

$$15\%\ of\ x = 60 \rightarrow 0.15x = 60 \rightarrow x = \frac{60}{0.15} = 400$$

25 percent of 400 is: $25\%\ of\ 200 = \frac{25}{100} \times 400 = 100$

2) Choice D is correct

$0.2 \times 4 = 0.8$, all options provided are equal to 0.8 except option D. $\frac{5}{15} \times 3 = 1$

3) Choice B is correct

Sara has N books. Mary has 5 more books than Sara. Then, Mary has $N + 5$ books. If Mary gives Sara 4 books, Mary will have: $N + 5 - 4 = N + 1$

4) Choice A is correct

If $\frac{3x}{2} = 30$, then $3x = 60 \rightarrow x = 20$

$\frac{2x}{5} = \frac{2\times20}{5} = \frac{40}{5} = 8$

5) Choice E is correct

$\frac{1}{5}$ of 40 is 8. Let's review the options provided:

(A) $0.3 \times 6 = 1.8$
(B) $0.3 \times 5 = 1.5$
(C) $0.2 \times 30 = 6$
(D) $0.2 \times 35 = 7$
(E) $0.2 \times 39.5 = 7.9$

Option E is the closest to 8.

6) Choice A is correct

The diagonal of the square is 6. Let x be the side.

Use Pythagorean Theorem: $a^2 + b^2 = c^2$

$x^2 + x^2 = 6^2 \Rightarrow 2x^2 = 6^2 \Rightarrow 2x^2 = 36 \Rightarrow x^2 = 18 \Rightarrow x = \sqrt{18}$

The area of the square is:

$\sqrt{18} \times \sqrt{18} = 18$

7) Choice B is correct

The sum of supplement angles is 180. Let x be that angle. Therefore, $x + 8x = 180$

$9x = 180$, divide both sides by 9: $x = 20$

8) Choice B is correct

Use the formula for Percent of Change

$\frac{\text{New Value} - \text{Old Value}}{\textit{Old Value}} \times 100\ \%$

$\frac{28-40}{40} \times 100\ \% = -30\ \%$ (negative sign here means that the new price is less than old price)

9) Choice B is correct

$x = 20 + 125 = 145$

10) Choice D is correct

The perimeter of the trapezoid is 54.

Therefore, the missing side (height) is $= 54 - 18 - 12 - 14 = 10$

Area of the trapezoid: $A = \frac{1}{2} h\ (b_1 + b_2) = \frac{1}{2}\ (10)\ (12 + 14) = 130$

11) Choice A is correct

If the score of Mia was 60, therefore the score of Ava is 30. Since, the score of Emma was half as that of Ava, therefore, the score of Emma is 15.

12) Choice D is correct

Let x be the number. Write the equation and solve for x.

$\frac{2}{3} \times 18 = \frac{2}{5}\ .\ x \Rightarrow \frac{2 \times 18}{3} = \frac{2x}{5}$, use cross multiplication to solve for x.

$5 \times 36 = 2x \times 3 \Rightarrow 180 = 6x \Rightarrow x = 30$

13) Choice D is correct

Let x be the number. Then: $3x + 6 = 30$

Solve for x: $3x + 6 = 30 \rightarrow 3x = 30 - 6 = 24 \rightarrow x = 24 \div 3 = 8$

14) Choice D is correct

Simplify and solve for x in the equation. $4(x + 1) = 6(x - 4) + 20 \rightarrow 4x + 4 = 6x - 24 + 20$

$4x + 4 = 6x - 4$

Subtract $4x$ from both sides: $4 = 2x - 4$, Add 4 to both sides: $8 = 2x$, $4 = x$

15) Choice E is correct

Five years ago, Amy was three times as old as Mike. Mike is 10 years now. Therefore, 5 years ago Mike was 5 years. Five years ago, Amy was: $A = 3 \times 5 = 15$

Now Amy is 20 years old: $15 + 5 = 20$

16) Choice B is correct

Let x be one-kilogram orange cost, then: $3x + (2 \times 4.2) = 26.4 \rightarrow 3x + 8.4 = 26.4 \rightarrow$ $3x = 26.4 - 8.4 \rightarrow 3x = 18 \rightarrow x = \frac{18}{3} = \$6$

17) Choice A is correct

average $= \frac{\text{sum of terms}}{\text{number of terms}}$

The sum of the weight of all girls is: $18 \times 60 = 1080$ kg

The sum of the weight of all boys is: $32 \times 62 = 1984$ kg

The sum of the weight of all students is: $1,080 + 1,984 = 3,064$ kg

Average $= \frac{3064}{50} = 61.28$

18) Choice C is correct

Solve for x in the equation.

$$6(x + 4) = 72 \rightarrow 6x + 24 = 72 \rightarrow 6x = 72 - 24 = 48 \rightarrow x = 48 \div 6 = 8$$

19) Choice C is correct

Let x be the number. Write the equation and solve for x.

$(24 - x) \div x = 3$

Multiply both sides by x.

$(24 - x) = 3x$, then add x both sides. $24 = 4x$, now divide both sides by 4. $x = 6$

20) Choice D is correct

Only option D is correct. $\frac{5}{6} = 0.83 \rightarrow 0.8 < \frac{5}{6}$

21) Choice D is correct

In a group of 5 books, the average number of pages is 24. Therefore, the sum of pages in all 5 books is (5 × 24 =) 120. Mary adds a book with 30 pages to the group. Then, the sum of pages in all 6 books is (5 × 24 + 30 =) 150. The new average number of pages per book is: $\frac{150}{6} = 25$

22) Choice B is correct

Choices A, C, D, and E are incorrect because 80% of each of the numbers is a non-whole number.

A. 49 $\quad 80\%\ of\ 49 = 0.80 \times 49 = 39.2$

B. 35 $\quad 80\%\ of\ 35 = 0.80 \times 35 = 28$

C. 32 $\quad 80\%\ of\ 32 = 0.80 \times 32 = 25.6$

D. 12 $\quad 80\%\ of\ 12 = 0.80 \times 12 = 9.6$

E. 8 $\quad 80\%\ of\ 8 = 0.80 \times 8 = 6.4$

23) Choice D is correct

$$\frac{2}{5} \times 25 = \frac{50}{5} = 10$$

24) Choice D is correct

The red box is 20% greater than the blue box. Let x be the capacity of the blue box. Then:

$$x + 20\%\ of\ x = 30 \rightarrow 1.2x = 30 \rightarrow x = \frac{30}{1.2} = 25$$

25) Choice A is correct

$One\ liter = 1000\ \text{cm}^3 \rightarrow 6\ liters = 6{,}000\ \text{cm}^3$

$6{,}000 = 15 \times 5 \times h \rightarrow h = \frac{6000}{75} = 80\ \text{cm}$

SSAT Middle Level Mathematics Practice Test 1

Section 2

1) Choice C is correct

$\$9 \times 10 = \$90$

Petrol use: $10 \times 2 = 20$ liters

Petrol cost: $20 \times \$1 = \$20$

Money earned: $\$90 - \$20 = \$70$

2) Choice C is correct

From the options provided, only C ($\frac{1}{6}$) is less than $\frac{1}{5}$.

3) Choice C is correct

Amy earns $30.00 per hour now. $30.00 per hour is 20 percent more than her previous rate. Let x be her rate before her raise. Then:

$$x + 0.20x = 30 \rightarrow 1.2x = 30 \rightarrow x = \frac{30}{1.2} = 25$$

John earns $26.40 per hour now. $26.40 per hour is 20 percent more than his previous rate. Let x be John's rate before his raise. Then:

$$x + 0.20x = 26.40 \rightarrow 1.2x = 26.40 \rightarrow x = \frac{26.40}{1.2} = 22$$

Amy earned $3.00 more per hour than John before their raises.

4) Choice D is correct.

Four people can paint 4 houses in 10 days. It means that for painting 8 houses in 10 days we need 8 people. To paint 8 houses in 5 days, 16 people are needed.

5) Choice A is correct.

$$N \times (5 - 3) = 12 \rightarrow N \times 2 = 12 \rightarrow N = 6$$

6) Choice D is correct.

The length of the rectangle is 18. Then, its width is 6.

$$18 \div 3 = 6$$

$Perimeter\ of\ a\ rectangle = 2 \times width + 2 \times length = 2 \times 6 + 2 \times 18 = 12 + 36 = 48$

7) Choice B is correct

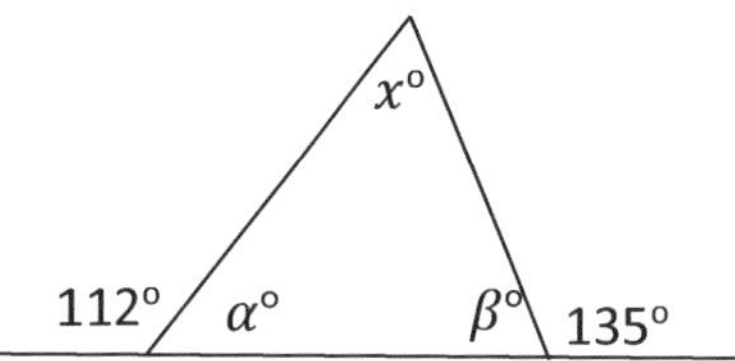

$\alpha = 180° - 112° = 68°$

$\beta = 180° - 135° = 45°$

The sum of all angles in a triangle is 180 degrees. Then:

$x + \alpha + \beta = 180° \rightarrow x = 180° - 68° - 45° = 67°$

8) Choice C is correct.

If $x \blacksquare y = 3x + y - 2$, Then: $4 \blacksquare 12 = 3(4) + 12 - 2 = 12 + 12 - 2 = 22$

9) Choice D is correct

The width of a rectangle is $4x$ and its length is $6x$. Therefore, the perimeter of the rectangle is $20x$. $Perimeter\ of\ a\ rectangle = 2(width + length) = 2(4x + 6x) = 2(10x) = 20x$

The perimeter of the rectangle is 80. Then: $20x = 80 \rightarrow x = 4$

10) Choice C is correct

The area of the floor is: 6 cm × 24 cm = 144 cm

The number is tiles needed = 144 ÷ 8 = 18

11) Choice B is correct

0.45 equals $45M$. Then: $45M = 0.45 \rightarrow M = \frac{0.45}{45} = 0.01$

12) Choice E is correct

$z = 3x + 6$, then, $2z = 2(3x + 6) = 6x + 12$

$$2z + 3 = 6x + 12 + 3 = 6x + 15$$

13) Choice D is correct.

$$20 = x \times 4 \rightarrow x = 20 \div 4 = 5$$

x equals to 5. Let's review the options provided:

A. $x + 4 \rightarrow 5 + 4 = 9$ 20 is not divisible by 9.

B. $2x - 4 \rightarrow 2 \times 5 - 4 = 6$ 20 is not divisible by 6.

C. $x - 2 \rightarrow 5 - 2 = 3$ 20 is not divisible by 3.

D. $x \times 4 \rightarrow 5 \times 4 = 20$ 20 is divisible by 20.

E. $x + 1 \rightarrow 5 + 1 = 6$ 20 is not divisible by 6.

The answer is D.

14) Choice C is correct

$\frac{1}{8} = 0.125 \rightarrow C = 5,$ $\frac{1}{20} = 0.05 \rightarrow D = 5 \rightarrow C \times D = 5 \times 5 = 25$

15) Choice C is correct

Use cross product to solve for x.

$$\frac{x}{x-3} = \frac{4}{5} \rightarrow 5 \times x = 4 \times (x-3) \rightarrow 5x = 4x - 12 \rightarrow x = -12$$

$$x - 5 = -12 - 5 = -17$$

16) Choice C is correct

x is the number of all sales profit and 2% of it is: $2\% \times x = 0.02x$

Employee's revenue: $0.2x + 7{,}000$

17) Choice B is correct

Number of biology book: 35, total number of books; $35 + 95 + 80 = 210$

the ratio of the number of biology books to the total number of books is: $\frac{35}{210} = \frac{1}{6}$

18) Choice C is correct.

$6{,}000 + A - 200 = 7{,}400 \rightarrow 6{,}000 + A = 7{,}400 + 200 = 7{,}600 \rightarrow A = 7{,}600 - 6{,}000 = 1{,}600$

19) Choice D is correct

Let x be all expenses, then $\frac{22}{100}x = \$660 \rightarrow x = \frac{100 \times \$660}{22} = \$3000$

He spent for his rent: $\frac{27}{100} \times \$3000 = \$810$

20) Choice C is correct

$5 \times M + 4 = 5 \rightarrow 5 \times M = 5 - 4 = 1 \rightarrow M = \frac{1}{5}$

21) Choice E is correct

$$\frac{42.6}{100} = 0.426$$

22) Choice B is correct

The angles on a straight line add up to 180 degrees. Then:

$x + 25 + y + 2x + y = 180$

Then, $3x + 2y = 180 - 25 \rightarrow 3(35) + 2y = 155$

$\rightarrow 2y = 155 - 105 = 50 \rightarrow y = 25$

23) Choice B is correct

let x be the number of gallons of water the container holds when it is full.

Then; $\frac{7}{24}x = 3.5 \rightarrow x = \frac{24 \times 3.5}{7} = 12$

24) Choice C is correct.

The ratio of lions to tigers is 5 to 3 at the zoo. Therefore, total number of lions and tigers must be divisible by 8.

$$5 + 3 = 8$$

From the numbers provided, only 98 is not divisible by 8.

25) Choice A is correct

If x is greater than 18, then $\frac{1}{6}$ of x must be greater than: $\frac{1}{6} \times 18 = 3$.

SSAT Middle Level Mathematics Practice Tests Explanations

SSAT Middle Level Mathematics Practice Test 2

Section 1

1) Choice C is correct

Use distance formula:

Distance = Rate × time ⇒ 420 = 50 × T, divide both sides by 50. 420 / 50 = T ⇒ T = 8.4 hours.

Change hours to minutes for the decimal part. 0.4 hours = 0.4 × 60 = 24 minutes.

2) Choice B is correct

To find the discount, multiply the number by (100% – rate of discount).

Therefore, for the first discount we get: (D) (100% – 20%) = (D) (0.80) = 0.80 D

For increase of 10 %: (0.85 D) (100% + 10%) = (0.85 D) (1.10) = 0.88 D = 88% of D or 0.88D

3) Choice A is correct

0.55 equals $55M$. Then: $0.55 = 55M \rightarrow M = \frac{0.55}{55} = 0.01$

4) Choice B is correct

Use simple interest formula:

$I = prt$ (I = interest, p = principal, r = rate, t = time)

t is for one year. For 3 months, t is ¼ or 0.25

$$I = (4{,}800)(0.05)(0.25) = 60$$

5) Choice A is correct

Three times a certain number, increased by 12, is equal to 39. Write an equation and solve.

$$3x + 12 = 39 \rightarrow 3x = 39 - 12 = 27 \rightarrow x = \frac{27}{3} = 9$$

6) Choice A is correct

30 percent of a number is 150. Therefore, the number is 500.

$$0.30x = 150 \rightarrow x = \frac{150}{0.30} = 500$$

12 percent of 500 is 60. $0.12 \times 500 = 60$

7) Choice E is correct

average = $\frac{\text{sum of terms}}{\text{number of terms}} \Rightarrow 18 = \frac{13+15+20+x}{4} \Rightarrow 72 = 48 + x \Rightarrow x = 24$

8) Choice C is correct

Add the first 5 numbers. 40 + 45 + 50 + 35 + 55 = 225

To find the distance traveled in the next 5 hours, multiply the average by number of hours.

Distance = Average × Rate = 50 × 5 = 250

Add both numbers. 250 + 225 = 475

9) Choice C is correct

John has N toy cars. Jack has 4 more cars than John. Therefore, Jack has N + 4 toy cars. Jack gives John 3 cars. Now, Jack has (N + 4 – 3) N + 1 toy cars.

10) Choice C is correct

$$\frac{x+4}{5} = 2 \rightarrow x + 4 = 2 \times 5 = 10 \rightarrow x = 10 - 4 = 6$$

11) Choice E is correct

Th ratio of boy to girls is 2:3. Therefore, there are 2 boys out of 5 students. To find the answer, first divide the total number of students by 5, then multiply the result by 2.

600 ÷ 5 = 120 ⇒ 120 × 2 = 240

12) Choice C is correct

Let x be the number. Write the equation and solve for x.

$\frac{2}{3} \times 18 = \frac{2}{5} \cdot x \Rightarrow \frac{2 \times 18}{3} = \frac{2x}{5}$, use cross multiplication to solve for x.

$5 \times 36 = 2x \times 3 \Rightarrow 180 = 6x \Rightarrow x = 30$

13) Choice A is correct

One pound of cheese costs \$0.96. $One\ pound = 16\ ounces$

16 ounces of cheese costs \$0.96. Then, 1 ounce of chees costs (0.96 ÷ 16) \$0.06.

6 ounces of cheese costs (6 × \$0.06) \$0.36.

14) Choice D is correct

Write the equation and solve for B:

$0.60A = 0.20B$, divide both sides by 0.20, then you will have $\frac{0.60}{0.20}A = B$, therefore:

$B = 3A$, and B is 3 times of A or it's 300% of A.

15) Choice B is correct

The question is this: 530.40 is what percent of 624?

Use percent formula:

part = $\frac{\text{percent}}{100}$ × whole

530.40 = $\frac{\text{percent}}{100}$ × 624 ⇒ 530.40 = $\frac{\text{percent} \times 624}{100}$ ⇒ 53040 = $\text{percent} \times 624 \Rightarrow$

percent = $\frac{53040}{624}$ = 85

530.40 is 85 % of 624. Therefore, the discount is: 100% – 85% = 15%

16) Choice D is correct

If 17 balls are removed from the bag at random, there will be one ball in the bag.

The probability of choosing a brown ball is 1 out of 18. Therefore, the probability of not choosing a brown ball is 17 out of 18 and the probability of having not a brown ball after removing 17 balls is the same.

17) Choice C is correct

Let x be the number. Write the equation and solve for x.

$(24 - x) \div x = 3$, Multiply both sides by x.

$(24 - x) = 3x$, then add x both sides. $24 = 4x$, now divide both sides by 4. $x = 6$

18) Choice A is correct

The percent of girls playing tennis is: $40\% \times 25\% = 0.40 \times 0.25 = 0.10 = 10\%$

19) Choice D is correct

The failing rate is 11 out of 55 = $\frac{11}{55}$

Change the fraction to percent:

$\frac{11}{55} \times 100\% = 20\%$

20 percent of students failed. Therefore, 80 percent of students passed the exam.

20) Choice E is correct

$$3x + 10 = 47 \rightarrow 3x = 47 - 10 = 39 \rightarrow x = \frac{39}{3} = 13$$

21) Choice A is correct

$N \times \frac{4}{3} \times 5 = 0$, then N must be 0.

22) Choice E is correct

$12 is what percent of $48?

$12 \div 48 = 0.25 = 25\%$

23) Choice C is correct

Let x be the number. Write the equation and solve for x.

$40\%\ of\ x = 4 \Rightarrow 0.40\,x = 4 \Rightarrow x = 4 \div 0.40 = 10$

24) Choice E is correct

$$\frac{z}{5} = 5 \rightarrow z = 5 \times 5 = 25$$

$$z + 3 = 25 + 3 = 28$$

25) Choice A is correct

Let x be the number of years. Therefore, \$2,000 per year equals $2000x$.

starting from \$24,000 annual salary means you should add that amount to $2000x$.

Income more than that is:

$$I > 2000x + 24000$$

SSAT Middle Level Mathematics Practice Test 2

Section 2

1) Choice C is correct

John has x dollars and he receives \$150. Therefore, he has $x + 150$.

He then buys a bicycle that costs \$120. Now, he has: $x + 150 - 120 = x + 30$

2) Choice A is correct

$$\frac{x-3}{8} + 4 = 20 \rightarrow \frac{x-3}{8} = 20 - 4 = 16 \rightarrow x - 3 = 16 \times 8 = 128 \rightarrow$$
$$x = 128 + 3 = 131$$

3) Choice E is correct

Jason needs an 75% average to pass for five exams. Therefore, the sum of 5 exams must be at least 5 × 75 = 375

The sum of 4 exams is:

68 + 72 + 85 + 90 = 315.

The minimum score Jason can earn on his fifth and final test to pass is: 375 – 315 = 60

4) Choice B is correct

The width of a rectangle is $6x$ and its length is $8x$. Then, the perimeter of the rectangle is $28x$.

$$Perimeter\ of\ a\ rectangle = 2(width + length) = 2(6x + 8x) = 28x$$

The perimeter of the rectangle is 56. Then: $28x = 56 \rightarrow x = 2$

5) Choice C is correct

Use simple interest formula: $I = prt$ (I = interest, p = principal, r = rate, t = time)

$$I = (8{,}000)(0.045)(5) = 1{,}800$$

6) Choice D is correct

$(9 - 5) \times 4 = 8 + \square$

Then: $4 \times 4 = 8 + \square$, $16 = 8 + \square$, then $\square = 8$

7) Choice C is correct

The distance between Jason and Joe is 9 miles. Jason running at 5.5 miles per hour and Joe is running at the speed of 7 miles per hour. Therefore, every hour the distance is 1.5 miles less.

9 ÷ 1.5 = 6

8) Choice E is correct

There are y tables that can each seat 5 people and there are x tables that can each seat 8 people. Therefore, $5y + 8x$ people can be seated in the classroom.

9) Choice E is correct

The formula for the area of the circle is: $A = \pi r^2$

The area of the circle is 64π. Therefore:

$A = \pi r^2 \Rightarrow 64\pi = \pi r^2$

Divide both sides by π: $64 = r^2 \Rightarrow r = 8$

Diameter of a circle is 2 x radius. Then: Diameter = 2 × 8 = 16

10) Choice D is correct

To find the discount, multiply the number by (100% – rate of discount).

Therefore, for the first discount we get: (200) (100% – 15%) = (200) (0.85) = 170

For the next 15 % discount: (200) (0.85) (0.85)

11) Choice D is correct

$4 \div \frac{1}{6} = 24$

12) Choice A is correct

The width of the rectangle is twice its length. Let x be the length. Then, $width = 2x$

Perimeter of the rectangle is 2 (width + length) = $2(2x + x) = 60 \Rightarrow 6x = 60 \Rightarrow x = 10$

Length of the rectangle is 10 meters.

13) Choice C is correct

$$3x + 12 = 48 \rightarrow 3x = 48 - 12 = 36 \rightarrow x = \frac{36}{3} = 12$$

14) Choice C is correct

average (mean) = $\frac{\text{sum of terms}}{\text{number of terms}} \Rightarrow 88 = \frac{\text{sum of terms}}{50} \Rightarrow$ sum = 88 × 50 = 4400

The difference of 94 and 69 is 25. Therefore, 25 should be subtracted from the sum.

4400 – 25 = 4375

mean = $\frac{\text{sum of terms}}{\text{number of terms}} \Rightarrow$ mean = $\frac{4375}{50}$ = 87.5

15) Choice D is correct

average = $\frac{\text{sum of terms}}{\text{number of terms}} \Rightarrow$ (average of 6 numbers) 12 = $\frac{\text{sum of numbers}}{6} \Rightarrow$sum of 6 numbers is

12 × 6 = 72

(average of 4 numbers) 10 = $\frac{\text{sum of numbers}}{4} \Rightarrow$sum of 4 numbers is 10 × 4 = 40

sum of 6 numbers – sum of 4 numbers = sum of 2 numbers

72 – 40 = 32 average of 2 numbers = $\frac{32}{2}$ = 16

16) Choice B is correct

$$x + 5 = 8 \rightarrow x = 8 - 5 = 3$$

$$2y - 1 = 5 \rightarrow 2y = 6 \rightarrow y = 3$$

$$xy + 10 = 3 \times 3 + 10 = 19$$

17) Choice B is correct

The deck contains 13 Hearts. Then, the probability of choosing a Hearts is $\frac{13}{52} = \frac{1}{4}$

18) Choice B is correct

From the options provided, only $\frac{1}{4}$ is greater than $\frac{1}{5}$.

19) Choice A is correct

2,500 out of 55,000 equals to $\frac{2500}{55000} = \frac{25}{550} = \frac{1}{22}$

20) Choice D is correct

$x = 9$, then $2x + 6 = 2 \times 9 + 6 = 18 + 6 = 24$

21) Choice C is correct

the population is increased by 15% and 20%. 15% increase changes the population to 115% of original population.

For the second increase, multiply the result by 120%.

(1.15) × (1.20) = 1.38 = 138%

38 percent of the population is increased after two years.

22) Choice A is correct

First, find the number. Let x be the number. Write the equation and solve for x.

150 % of a number is 75, then: $1.5 \times x = 75 \Rightarrow x = 75 \div 1.5 = 50$

90 % of 50 is: 0.9 × 50 = 45

23) Choice B is correct

Plug in 104 for F and then solve for C.

$$C = \frac{5}{9}(F - 32) \Rightarrow C = \frac{5}{9}(104 - 32) \Rightarrow C = \frac{5}{9}(72) = 40$$

24) Choice B is correct

The perimeter of the trapezoid is 36.

Therefore, the missing side (height) is = 36 – 8 – 12 – 6 = 10

Area of a trapezoid: $A = \frac{1}{2} h (b1 + b2) = \frac{1}{2} (10) (6 + 8) = 70$

25) Choice D is correct

If the length of the box is 27, then the width of the box is one third of it, 9, and the height of the box is 3 (one third of the width). The volume of the box is:

$$Volume\ of\ a\ box = (length) \times (width) \times (height) = (27) \times (9) \times (3) = 729$$

"Effortless Math Education" Publications

Effortless Math Education authors' team strives to prepare and publish the best quality Mathematics learning resources to make learning Math easier for all. We hope that our publications help you or your student learn Math in an effective way.

We all in Effortless Math wish you good luck and successful studies!

Effortless Math Authors

www.EffortlessMath.com

... So Much More Online!

- ✓ FREE Math lessons
- ✓ More Math learning books!
- ✓ Mathematics Worksheets
- ✓ Online Math Tutors

Need a PDF version of this book?

Visit www.EffortlessMath.com

Or send email to: info@EffortlessMath.com